STEM books by Dr. Wood

- Coral Reefs, ISBN: 979-8-9890262-1-0
- OREOs & Milk, ISBN: 979-8-9890262-2-7
- Hurricane Dorian and the FEMA House, ISBN: 979-8-9890262-3-4
- Treasure Trove! ISBN: 979-8-9890262-4-1
- Radioactive! ISBN: 979-8-9890262-5-8
- Whales!!!! ISBN: 979-8-9890262-6-5

Textbooks by Dr. Wood

- Underwater Robotics. ISBN-13: 978-1367883208
- Underwater Archaeology Engineering Bible Technology & Practice. ISBN: 978-1-60797-996-8
- Hydro-acoustics for Undergrads
- Surface Treatments for Dropwise Condensation Applications

Acknowledgments

I wish to acknowledge the young adults who spent valuable time critiquing the *Save the World Kids* series of books: Suzanne Dixon, Parker Baillon, Stephen Coster, Christine Coster, Gabor Papp, and my wife, Gisela Suzanne-Bahr Wood, who inspired me to start writing these novels.

Save the World!

This series of novels is about a family that tackles various societal issues: Father Sam Jones, age 39; Mother Suzanne Jones, age 35; daughter Diane, age 11; son Ben, age 6; and dog Freddy, a 55-pound female golden retriever. This book includes deep underwater caves, aquifers, shipwrecks, drug smugglers, and hurricanes. It covers the design of an autonomous underwater vehicle, called an AUV, and how middle school students can collaborate with college students to do great things. All ocean engineering undergraduate and graduate students are real, although they may not have worked on the OREO project.

OREOS & MILK

Save The World Series - Ocean Engineering (a STEM book).

Only 6.7% of women graduate with a STEM degree. When we encourage girls to pursue STEM, we double our innovation potential.

STEM books invite thinking by

- Modeling Real-World Innovation
- Embracing Real-World Design, Invention, and Innovation
- Connecting with Authentic Experiences
- Showing Assimilation of New Ideas
- Illustrating Teamwork, Diverse Skills, Creativity, and Cooperation
- Inviting Divergent Thinking and Doing
- Integrating Interdisciplinary and Creative Approaches
- Exploring Multiple Solutions to Problems
- Addressing Connections between STEM disciplines
- Exploring *Engineering Habits of Mind*
    - Systems thinking
    - Creativity
    - Optimization
    - Collaboration

STEM books represent the practices of science and engineering by

- Asking questions, solving problems, designing, and redesigning
- Integrating STEM disciplines
- Showing the progressive changes that characterize invention and engineering by:
    - Demonstrating designing, redesigning, improving, building, or repairing a product or idea.
    - Showing the process of working through trial and error
    - Progressively developing better engineering solutions
    - Analyzing efforts and making necessary modifications along the way.
- Illustrates that failure might happen, which is acceptable, providing reflection and learning.

Cataloging-in-Publication data on file with the Library of Congress

ISBN: 979-8-9890262-2-7

eBook ISBN: 979-8-9890262-0-3

Audio Book ISBN: 979-8-9890262-9-6

Published and Printed in the United States of America

Ocean Engineering

A multidisciplinary field of technology applied to the ocean environment that integrates oceanography, materials science, mechanical, civil, computer, software, marine, chemical, electrical, electronic engineering disciplines, naval architecture, and applied physics.

Ocean engineering is a rapidly growing and dynamic field with opportunities that are expanding as people turn to the oceans for food, transportation, and energy. One of the great things about ocean engineering is that many different types of engineers can work together to find solutions for ocean infrastructure, research, and utilization.

In addition to creating ROVs, ocean engineers develop underwater structures, oil rigs, and wave buoys for data collection, and they are hard at work developing ways to capture the energy of waves and turn it into electricity. They develop transportation systems, plan new uses for waterways, design deep-water ports, and integrate land and water transportation systems and methods. They are concerned with discovering, producing, and transporting offshore oil and developing new ways to protect marine wildlife and beaches against the unwanted consequences of offshore oil production and storm erosion.

Ocean engineers study all aspects of the ocean environment to determine the influence of people on the oceans and the effects of the ocean on ships and other marine vehicles and structures. Their work is global and has never been more important because these professions connect people and places in a way unmatched by other engineering careers. Think of Earth as one big web of biodiversity connecting us to all living things. Many may say that the Pacific Ocean separates the United States and Japan. However, in the maritime industry, they say that the oceans connect them.

Table of Contents

PREFACE 1

THE BAHAMAS 5

THE TEAM 12

INTO THE DEPTHS 27

HOT CHICKS - COOL DUDES 34

DANCE THE NIGHT AWAY 43

ROCKETS TO THE MOON 53

CHECKLISTS 58

SEAWORLD 70

DRUGS, SUBS, AND BUGS 81

POOL TESTING 85

LAKE TESTING 89

WES SKILES PEACOCK SPRING STATE PARK 93

THE SPIEGEL GROVE SHIPWRECK 101

SPIEGEL GROVE DIVE DAY 107

DEAD BODY! 113

SEARCH FOR TREASURE 118

MISSING! 126

FLIGHT TO THE BAHAMAS 129

BLUEHOLE 137

POSSIBILITIES 149

FOLLOW THE UNKNOWN TRACK 164

THE CHASE 170

THE CAPTURE 177

HURRICANE DORIAN 182

PREFACE

Note from the author: With a few exceptions (Thomas, Dean, Zack, Sergeant Mitchell, Sam, Ben, Svetlana, and the drug smugglers), the people in this book are based on real people. None of the locations are fictional. Only the drug smuggling story about importing drugs from Abaco Island to Florida is fictional. All the other stories are true. The OREO AUV exists and is being built for cave and shipwreck exploration at the Underwater Technology Laboratory at the Florida Institute of Technology in Melbourne, Florida.

Characters

Diane Jones is an energetic twelve-year-old girl with an extreme interest in everything electronic, especially robots. She also has a deep passion for the oceans and marine life. She has had visions of being a dolphin trainer or a rescuer of manatees or sea turtles. Her dad, Sam, is a mechanical engineer with his own company designing robot systems for various industries. He suggested to Diane that she investigate the ocean engineering profession to combine her passions of electronics, mechanics, and marine biology.

Thomas O'Conner and **Dean Bond**, 12 years old, are Diane's techie friends. Whenever something comes up in the tech world, these two jump in and learn everything they can about it.

Zack Fernwood, Diane's friend - Diane met Zack, 15, a few months ago during her investigation of global warming, sea level rise, and the impacts these two phenomena have on coral reefs off the coast of Florida. Zack has lived in the Florida Keys his whole life, and even though he is only twelve, he can scuba dive and run any motorboat in the Florida Keys. His dad, a ship marine engineer, taught Zack everything about motors, pumps, and what it takes to care for and repair small boats.

Sam Jones, Diane's dad, a professional mechanical engineer, specializes in designing robot systems. He is skilled in mechanical, electrical, and software engineering, specifically in artificial intelligence, control systems, and navigation. His company designs equipment for various industries, including the NSA, Homeland Security, and NASA.

Suzanne Jones, Diane's mom, a computer science professor, focuses on large data mining. She excels at solving complicated problems with innovative solutions. A certified scuba diver, Suzanne has enjoyed the ocean since childhood. She has taken a five-year leave of absence to work on a few projects while raising a family.

Ben Jones, 8, Diane's brother, is in the second grade and always follows Diane around and gets in the way. Luckily for Diane, Ben gets bored quickly and ends up playing with his toys while Diane does her investigations on the computer or her electronics desk, where she has just about every robot mechanism on the market.

Dr. Wood, Professor and Ocean Engineering Program Chair at the Florida Institute of Technology is the head of the Underwater Technology Laboratory, where undergraduate and graduate students conceive, design, build, and test autonomous and remotely operated systems for the ocean. Every summer, he takes his students on multiday cruises on the Florida Institute of Oceanography's research vessel *Hogarth* to the Florida Keys and the Dry Tortugas.

Brooklyn, a 22-year-old graduate student in Ocean Engineering, finished her undergraduate degree at the Florida Institute of Technology in Ocean Engineering and is continuing to earn her master's degree. Her thesis is on autonomous underwater vehicles to map caves.

Tij, a Ph.D. Ocean Engineering graduate student from India is skilled at modeling the actions of any vehicle in or on the water. He led the team in building the Nautilus, a vehicle that can fly or swim.

Brian, a Ph.D. Ocean Engineering graduate student from Tennessee designed an acoustic identification system to detect, identify, and target any acoustic sound in the ocean. His system can tell you if the sound is Bill's shrimp boat, Peggy's fishing boat, a blue whale from the Pacific, a bottlenose dolphin from the Atlantic, or an enemy submarine passing by just by listening to the sounds of a ship's propeller or other marine sounds.

Cody, a sophomore undergraduate student from a small town in central Florida, has been interested in the oceans since he was young. He joined Dr. Wood's team the moment he arrived at the university. He jumped at the opportunity to become machine shop and 3-D printer certified. It was during this training that he met Derek.

Derek, a college sophomore from a small town in the Pacific Northwest, jumped at the opportunity to join Dr. Wood's team as a work-study student. He had initially signed up for mechanical engineering but switched after meeting Cody, who had told him about the projects underway in Dr. Wood's laboratory.

Doug, an engineer, has been a design and test engineer of systems designed for the ocean for more than thirty years. His experience on the sea and in the laboratory is second to none. Taking apart and repairing a SideScan SONAR, current meter, or REMUS autonomous underwater vehicle is what he calls fun. With experience with naval acoustical systems, deep-sea diving (to 300 feet depth), and underwater archaeology in the Netherlands and Sweden, Doug's abilities rival those in the best institutes worldwide.

Svetlana, an adventurer and archaeologist, was educated in the Ukraine in the small city of Mykolaiv just east of Odesa on the Black Sea. Her base studies were in naval architecture, but when she emigrated to the United States, she applied to a university in Texas to become an underwater archaeologist. After completing her degree, she worked on underwater projects throughout the world.

Arthur, a Government Official, is good friends with Dr. Wood; Arthur received a graduate degree from the University of Miami, where he was known for his Calypso music. Years later, Arthur and Dr. Wood meet during this adventure.

Sergeant Mitchell, a very competent police officer who tends to receive all the out-of-the-ordinary assignments. Good-hearted with young people, he volunteers to assist in the search for the lost cave divers.

Rob and Leanne, Dr. Wood's friends from Saskatchewan, Canada, have known Dr. Wood since they were in graduate school at the University of Miami.

Mr. Rodrigues, a drug smuggler, is in command of the 'Bahama Mamas.' Mr. Rodrigues is a 'drug cartel lord' with links to a Colombian cartel and has been smuggling drugs into the US for several years.

Miguel, a drug smuggler, second in command of the 'Bahama Mamas,' is in a drug cartel associated with a Colombian cartel and has been smuggling drugs into the US for several years with Mr. Rodrigues.

Carlos, drug smuggler, Mr. Rodrigues' number one henchman.

Luke is the pilot of a drug smuggler submarine.

THE BAHAMAS

"You went ashore! You are lucky you didn't get shot!" exclaimed Randy, a fellow sailboat owner.

"Wait! What? This is the Bahamas!" Leanne said in confusion. Leanne had been telling Randy about how they noticed the new development in Crab Cay.

Typically, tourists think of the Bahamas for its beautiful beaches, warm crystal-clear waters, friendly hospitality, and the sailing that can be done in the area. Scuba tourists think of the incredible coral reefs and the numerous wrecks they can explore. Then there is the adventure diver, who likes to enter caves. Like many other islands, the Bahamas is composed primarily of limestone. The Bahama Plateau was affected by sea-level changes during the Pleistocene Ice Age, which ended 11,700 years ago. During this time, the change in water level encouraged the development of the caves. So, the Bahamas has numerous caves and sinkholes, especially Abaco Island. Because of their principal coloration, these caves are

BLUE HOLES

Blue holes get their name from the blue color of their surface water. This is generally the reflection of the sky on the water when viewed from an aircraft. The surface water in some blue holes can be dark, tannin, or even muddy, but the deep water below is usually very clear.

The Abaco Marine or ocean blue holes have entrances, which are often accessed from below sea level. These caves can be connected to each other and are affected by tides and currents.

known as 'Blue Holes.' These 'holes' are common in shallow waters, such as 'The Guardian Blue Hole System' or the 'Bight of Old Robinson' near Little Harbor, which has several openings in its shallows that are accessible by boat and make for exciting snorkeling for those who are not cave divers. Diving these holes is only for the most experienced divers.

ψ

Rob and Leanne's lifetime dream was to sail around the Caribbean, and scuba dive the famous blue holes of the Bahamas. These two adventurers were not of the adventuring type, but they liked the outdoors. Rob and Leanne were from Saskatchewan, Canada. They had rented a sailboat out of Freeport and spent their first week exploring. At one of their anchoring spots, they met a couple who told them about a blue hole, which had been wonderful on Little Abaco Island but seemed to have been destroyed by developers. Heeding the warning signs of 'NO TRESPASSING,' the couple did not investigate further.

That was enough for Rob and Leanne; they had to investigate, so they headed there the next morning. They anchored in Crab Cay for several nights near where the blue hole should be. On their first night, they noticed no sand on the beach, and everything was covered in fine silt underwater. The jellyfish were so thick that it was almost impossible to be in the water. Jellyfish populations are known to bloom in unhealthy ecosystems; in this case, it certainly felt that way. The place had changed. It wasn't until they paddled around in their dinghy the next day that they started to put the pieces together. The people they had talked to had visited the area years earlier and had swum a blue hole. Rob and Leanne went off searching for the blue hole but were confused. They had been informed that the blue hole was in the mangroves in a creek at the corner of the anchorage, and it did not look right. They kept looking until they found a small but distinctly dredged channel, which you could only see from head-on. Rob and Leanne dinghied in and saw an array of dredged canals and vegetation. They could not find the mangroves because they were no longer there. It looked like there would be a huge development of some sort. A marina? Condos with slips? But why? The Green Turtle Hotel resort a few miles away could not keep its spaces full, so why build a new harbor facility in an even more remote and not necessarily picturesque location, which is remote by land and water? It didn't make sense.

Further, they found earth-moving equipment which had not been used for quite some time. Perhaps this was another project out of funds, leaving it to the land to do her best to heal the damage. They looked for the blue hole without luck but found a tower that was placed to see the work site and a couple of old derelict mobile homes. Rob and Leanne climbed up and took a quick peek. What they saw made them sad: all the destruction and decimation of the mangroves. Not knowing what was happening, they decided to head back to their boat and then to a local harbor to ask questions about what happened at Crab Cay.

Randy proceeded to fill Leanne in on the story. "It is not a condo project. The Bahamas government approved a deal in which the Export-Import Bank of China would back 41 million dollars to a Chinese company, CHEC, to build a state-of-the-art commercial shipping port in the northern Abaco Island. A friend of mine, Bob, his boat is usually docked at the end of this pier. When Bob poked around, he discovered that the Chinese company had displayed a strong security presence. Here, look at some of these photos."

He showed Leanne some photos that he had taken. One with armed security and a big flag saying China Harbor. Not exactly what they had seen.

Leanne asked. "But why? It doesn't make sense. There are only two settlements close to Cooperstown and Foxtown. A large port is not viable."

Randy continued. "A different Chinese company had already built what could be the world's largest shipping port in Freeport, Bahamas. We all thought that this development would be used to facilitate the export of seafood from the Bahamas to China. People had previously run into suspected poaching fleets off the banks, and last year, a whole Chinese fleet of lobster boats was confiscated for illegal fishing in Bahamian waters. This brought some sense to the project, even though all of this is just wild speculation on our part."

"Interesting," Leanne said thoughtfully. "But why did they stop construction? Did the funding fail?"

"Nope," Randy continued. "Get this. It turns out that they came across an unexpected blue hole during the excavation. Yeah, that secret one that divers have been enjoying for decades. When researching the area, they found that a complete environmental study still needed to be completed before starting dredging. Therefore, the project is on hold while an environmental study can be conducted."

"Yes, I can understand that," said Leanne. "Blue holes are fragile ecosystems, and they are linked to underwater caverns. You cannot just fill one in. Especially since there are organisms found only in Bahamian blue holes."

Later that night, Rob and Leanne discussed the hole they wanted to dive into. Sadly, they doubted that it would stop a multimillion-dollar project.

"Leanne, this is probably the last opportunity to dive the hole. With construction at the Crab Cay site on hold and the company's primary construction site further down the coast where the actual port will be, no one will notice if we slip in for an extensive dive into the caves," said Rob.

"Rob, before we dive this weekend, let's find out more to ensure it is okay to dive there. We don't want to get in trouble.".

"Yes, but of course," replied Rob with a kiss.

Curious, Rob and Leanne kept asking around. They started hearing reports that a 60-foot-deep channel would be dredged for the new port.

That night, they continued the discussion.

"Leanne, this doesn't make sense. Marsh Harbor is less than 20 feet deep. Why 60 feet of water?" Rob asked.

Leanne had read an article in the local newspaper about the project, "Look here Rob, this sounds very positive, as if it is a well-supported endeavor, but the feedback the newspaper received from the locals was anything but. It says here that some locals were sad and felt like their entire nation was being sold, others knew little about the construction, and some were upset to hear that the mangroves were being removed."

"Well, sorry, they did. They are all gone," said Rob.

"The newspaper reported that although many jobs were promised, the wage offered by CHEC was so low that the Bahamas refused, and the paper also stated that the company's workers were not leaving the construction compound."

A day later, Rob and Leanne talked to two local business owners and told them about the rumors of where they were dredging, how deep it was, and how it didn't make sense.

They said, "Oh, but it does. Let me tell you why…" They explained that not too long ago, the Bahamian government approved a deal for a Chinese naval base on Cat Island just east of Exuma Island in the Bahamas. The United States got wind of it and let the Bahamas know it would not be in their best interest to proceed with this agreement, so the deal fell through.

Rob and Leanne then asked what boat needed that much water. The answer is simple: something either very big or needs to go deep.

That night, Rob and Leanne had friends over for dinner on the back of their boat. The discussion was about the blue hole and how crazy it was.

To make a long story short, the couple knew many things about the site. A friend of theirs who runs a survey boat was able to connect some of the dots of their activities. "The deal involves not only the container port, which will be located just south of where they saw the construction, more towards Cooperstown and the small bridge connecting Great Abaco with Little Abaco Island. Unfortunately, a tiny, filled-in land bridge blocks the flow between the islands. The Bohemian government plans to remove the old bridge and restore the tidal flow. Unfortunately, they are replacing it with a 150-foot-wide bridge. Again, huh? There is barely a creek between the two islands. What boats are going through there? Not many without a whole lot more dredging."

Rob immediately said to Leanne: "I'm glad we didn't wait another ten years. I'm not sure what to expect when we return there."

This sad discovery led them to want to stay and soak up as much of the northern Abaco as possible; it might be their last chance before the good fishing grounds are dredged away, causing the pristine beaches to erode, as they did in Crab Cay. They felt helpless for the blue holes and mangroves (the nurseries and filters of the sea) without knowing how to help.

A couple of days later, back at Crab Cay, Rob moored their small dive boat as close to the blue hole as possible. The dense mangroves were gone, just a ragged channel that cut into the edge of the blue hole. Once there, Rob and Leanne dropped into some unusually chilly water. The Bahamas is known for its warm water temperatures, often near 90 degrees in some of the marine caves here in the summertime, but this cave had an unusual upwelling of cold water from the depths, chilled to an icy 60 degrees! Something about the cave must have changed for such cold waters to flow from the depths.

Once Rob and Leanne got over the chill, they dropped into the vertical fissure where the cold, clear water emerged. A few small red snappers dashed away from the sight and sound of the noisy, bubble-generating intruders.

A very convenient tie-off at 20 feet that someone had put in made for a quick start to the dive, at least until Rob and Leanne got to the bottom of the fracture. There was one small, odd-shaped opening with a flake of rock sticking right into the center, leaving Rob and Leanne with no choice but to remove one of their two air tanks to get through. They tied it off to the flake, pulled out some slack, locked off the reel, and then dropped it down through the small hole. Following one of their removed air tanks, they squeezed through the entrance without much hassle, landing on a mangrove leaf-covered floor at a depth of 45 feet.

A small oval passage with clear water led away. The passage was only 4 to 5 feet in diameter and covered in colorful filter-feeding animals that gathered food from the reversing currents flowing through the cave. White, blue, brown, and orange sponges were scattered on the walls and ceiling, while the floor was a jumble of light gray organic mud, reddish-colored rock, and loosely layered mangrove tree leaves.

The passage continued southeast, slightly dipping to depths between 55 and 60 feet. There were several tight squeezes through the mud and eroded formations along the way, but each time, the passage opened again on the other side of the restrictions. Eventually, the passage began to become smaller, and slowly, the ceiling came closer and closer to the bottom. Knowing that they would be coming back through each of the odd-shaped restrictions in zero visibility, they decided to stop the dive with 25 feet of line left on the reel. Rob and Leanne tied off to one of the only solid wall projections they could find, cut the reel free, and dropped a line arrow at the end, 695 feet from where they had entered the cave.

They had nearly zero visibility during the first 4 or 5 minutes of the swim, making the odd-shaped restrictions challenging. Still, the water cleared, giving them plenty of opportunities to investigate a half-dozen side passages they had seen on the way in. They had to tie off each time they explored a side passage since the tide had changed and water was flowing into the cave. The passage Rob picked seemed to be paralleling the original passage, but they had to stop exploring when the line on the reel ran out.

They were beginning to feel chilled from the cold water, so they decided to head back and check out the different entrance openings at the front of the cave. After passing through the original entrance restriction, they could easily see another passage with light at the end off to the right. A short 30-foot swim through this tunnel, and they reached a vertical, 4-foot diameter opening where dozens of small snappers were fighting to stay in place in the strongly siphoning current.

Rob turned to Leanne, who had her arms crossed and was rubbing her upper arms with her hands as though she was trying to warm herself, signaling she was very cold. Rob nodded and stayed very close while they pulled into the sunlight and back into the small shallow bay to their boat. Their fingers were stiff from the cold water, and Leanne was shivering. It had been a fantastic dive! Tomorrow, they would explore further, as long as they didn't meet the Chinese company . . .

THE TEAM

Diane's friend Thomas, who became acquainted with Dr. Wood while he was doing a homework project on pollution in our oceans, informed her and his best friend, Dean, of a project he is involved in an advanced underwater submarine project where they are developing a new autonomous underwater vehicle (AUV) that does much more than what current AUVs do.

"Thomas," asked Diane, "I have two questions. What is an AUV? And who is this, Dr. Wood? I mean, is this legit? You are only in middle school; how did you get involved in this project?"

"Diane, this is so cool," said Thomas. "First, Dr. Wood used some of my data in one of his books, so we became friends. During one of our talks, Dr. Wood introduced me to two of his undergraduate students, Cody and Derek, and one of his graduate students, Brooklyn, who are studying ocean engineering at his university in Florida."

"Which university and what is ocean engineering?" Diane interrupted.

"Florida Institute of Technology in Melbourne, Florida, and I have no idea what ocean engineering is except they design things for the oceans, boats, subs, and stuff."

"OK, we can look it up later," Diane said.

"Exactly. Now, what the project is about is that there are huge caves in Florida and the Bahamas that no one has explored. Very deep and very long caves that scientists have been trying to explore for years, but no one has been able to do it. He is having his students make a vehicle to explore the caves, and they have invited me to join."

"Your joking," said Dean.

"No, he asked if I had any friends who would also be interested in joining. I'm going to call in on Zoom this Monday at 10 a.m. Are you two interested in joining the meeting?" Thomas asked.

"Are you kidding? Of course!" was their response.

"Tell us all about it," said Diane.

"The vehicle they envision is an autonomous underwater vehicle. They are starting with a remotely operated vehicle, known as an ROV. They will modify it by introducing a computer brain so the vehicle can act like a human controls it. What is really interesting is that the vehicle will be powered by the corrosion of aluminum (in other words, the oxidation of aluminum); can you believe that sticking aluminum in salt water can power an underwater motor?"

"OK, sounds interesting," said Dean, a little bit overwhelmed.

"Dr. Wood had his students design the vehicle: frame, the layout of where everything will be, and what computers and electronics are necessary.

"Cody has located new motors, a pressure chamber, and several other items from BlueRobotics. The Zoom on Monday will discuss the basic layout of the vehicle. What is known is that it must be quite small to fit in a cave or shipwreck and be able to run for days."

Thomas continued. "During the call, we will discuss the basic design of the vehicle, the computer design, and how to test the system and the integration of parts."

ψ

Monday, 10 a.m.

The Zoom connection was made just as Thomas said: "Remember that cartoon where all the sharks had lasers on their heads? POW, POW, POW, ZOWEE!"

"Yes, but come on," replied Diane, "act a little more grown up."

"Hmmm," Cody said out of the blue. "I don't know what you are talking about, but lasers on the vehicle are not a bad idea. We could have six lasers. Each on a pan tilt mechanism. If each mechanism rotated a known pattern and if we knew our velocity, we could map wherever we go using the lasers! Great idea, Thomas!"

Derek gave Cody an attaboy punch in the arm.

"With that introduction, Thomas, who have you invited to our Zoom?" Brooklyn asked.

"Hi everyone; this is Diane, and this is Dean. They are my two best friends. Diane is a genius at everything technical, computers, and stuff. This is Dean. He is a dunce. Just kidding. He may be quiet, but he really knows his way around a keyboard."

Brooklyn started. "Nice to meet you. I am Brooklyn, in charge of this project, and this is Cody and Derek, my two technical wingmen. A couple of others behind the scenes are also developing computer code for the vehicle: Anthony, Melanie, and Amanda. They are all graduate students who will be extremely busy this summer. So, beyond helping with the navigation and control code, I doubt we will see much of them. Diane and Dean, has Thomas filled you in on the project?"

They both jumped in, said yes, and immediately explained what Thomas had told them.

"Excellent! You both seem very interested. Perfect, so yes, we will design, build, and test this vehicle over the next few months. I talked to Dr. Wood, the professor in charge. We want to invite you to come to Florida this summer for the final construction and testing of the vehicle."

Diane, Thomas, and Dean stared in amazement.

Brooklyn continued. "We shall discuss that later. Today, we need to get into the details of the design. Thomas, Diane, and Dean fill in and comment on anything. Just because you are new to the project does not mean that you can't contribute. The main point is that the vehicle will go into caves and shipwrecks. So, it needs to be small and have no protrusions that can catch on to anything. Several things make up the vehicle: pressure chambers to hold electronics, motors, sensors, flotation, computers, etc. I am working on a list that we will add to as we go. And yes, lasers are an option, as are acoustic transducers, which I will tell you about later. Since you have been talking about lasers, I give you, Thomas, Dean, and Diane, the task of specifying one for the vehicle."

While Brooklyn continued explaining what was needed for the vehicle, Diana punched a few keys on her keyboard and exclaimed: "Hey everyone, I just did a quick search on the robot store's website. They have laser positioning devices for $120 each. It says here that they are great for knowing distances to objects. They even have a video showing a drone flying over cars. The drone maintains an exact distance above each automobile as it flies over them."

Brooklyn responded, "Well done, Diane! Now, this is teamwork. Email me that link so I can check out the specs and find four more from different places. When we determine the best, I will order six if they are small and accurate enough."

"A vehicle with lasers! How cool is that!" Said Thomas.

Brooklyn continued: "OK, I need to meet with Dr. Wood in a few minutes, so I would like to schedule another Zoom in four days. Same time. Can everyone make it?"

Everyone agreed.

"Excellent; I want everyone to look up various components we could integrate into the vehicle. Everyone, I will inform Dr. Wood about our new team members. Sorry, I must rush, but I cannot keep the boss waiting. See you all in four days."

ψ

Friday, 10 a.m.

Brooklyn returned everyone to task after making the Zoom link. "Now, let's talk about the vehicle. What thoughts do you have? Diane, you are first."

Diane started with a few other ideas. "I was thinking that we should have multiple GoPro cameras. One in each corner and one on each side. Let's see, at the going rate of $275 each plus waterproof housings at $50 apiece, it all comes to about $2,000."

"Wow," said Brooklyn. "I'll have to ask Dr. Wood if we can spend so much on cameras."

Diane continued. "Also, ask Dr. Wood for $50,000 for the software combining GoPro videos into a virtual reality presentation. The video we will get will be so cool!"

Dean, who had been left out of most of the conversation so far, piped up. "Do you mean that with these cameras, a person using virtual reality glasses can view the scene as if they were actually there? And no matter where you look, you will see everything: up, down, left, right? No way! Now that is so - wow!"

Dean continued: "This vehicle will be so super cool."

Everyone had big smiles on their faces. Brooklyn frowned. "I will ask, but I'm not sure he has that much money."

Cody and Derek came next, describing how they would need forward-backward and up-down thrusters in a weighted tilt mechanism to tilt the vehicle upward or downward. They also talked about needing small articulated robotic arms to grip small things.

"Excellent," concluded Brooklyn, "We need to do another Zoom. Diane, we need you to investigate the lasers further. Especially the integration of them using Arduino UNO or MEGA boards."

"No problem. I will get you a full report on them, and Thomas and Derek here can help. We also thought we might want to investigate acoustic sensors for when the water is too murky for the lasers to penetrate," said Diane.

"Great idea Diane," Brooklyn replied. "Cody and Derek. Could you investigate the aluminum oxide batteries that L3Harris is developing? I saw somewhere in the news that the BAE System's Riptide autonomous underwater vehicle plans to use this for their power source,"

Cody and Derek nodded, and Brooklyn continued: "I will work on the vehicle's layout with what we know. I will also investigate the electrical side of things, such as other sensors we need. I will email you the list that I have so far. Any questions? Next week."

Everyone said, "Bye."

Diane, Thomas, and Dean looked at each other and grinned.

Thomas said, "Now, this is really cool. My mom and dad will be tickled pink that we are part of a team to build an autonomous underwater vehicle to track what bad people are putting in our water supply, mapping uncharted deep-water caves, or investigating sunken shipwrecks."

ψ

Zoom – Friday 10 a.m. - one week later

"Oh, everyone." Brooklyn started in. "I need you all to install Fusion-360 on your computers, and after giving you each permission, you will all be able to see the computer-aided design (CAD) drawings I am working on. We also need a few Zoom sessions to get everyone up to speed using the CAD software."

"Sweet," Cody replied, "This fall, we will be ahead of everyone else in our ocean engineering computer application class. Excellent!"

"Ah, Brooklyn. I am a little lost here," Dean said, "I need to familiarize myself with CAD. Can you explain a little more about it?"

"Sure Dean," Brooklyn continued. She always loved to teach others about CAD. She showed everyone some of her design concepts.

"It is always best to first sketch your ideas. Try to come up with some possibilities. I try to have five different concepts for each idea. Then, I make lists of the positives and negatives of each. As a team, we need to evaluate each one and either select one or come up with another best solution that we all derived together."

They all looked at her sketches.

"Brooklyn," said Diane. "Great drawings. I just love your overall concept. It looks like an OREO cookie. How big is it?"

"About two feet in diameter and a little over a foot high. I think it will weigh about 50 pounds when finished," said Brooklyn.

"Hmm," continued Diane. "You know, if you added connection points here, here, and here, you could make a backpack carrier for it. It would be more than I could carry, but I am sure a strong guy could carry it alone."

"Cool idea," said Dean. "That's too heavy for me, but my older brother often carries over 60 pounds when hiking in Yosemite, California."

"Perfect! Diane, I would like you to take these drawings and make modifications to incorporate a backpack-carrying concept. That will be very handy if we carry it across rough terrain," Brooklyn said.

The team discussed the physical design for another half hour. Brooklyn then broached the topic of what everyone knew about electronics. She was met with blank stares. Diane started slowly.

"Brooklyn, none of us has had much experience with electronics. My dad and I have been playing with microcontrollers called the Arduino board, the Raspberry Pi, and all types of sensors. We got them a month ago, but he has been so busy at work that we have not progressed very far," said Diane.

"Fantastic!" Brooklyn said emphatically. "You are way ahead of many in college! The three of you get familiar with all the sensors over the next few weeks. Go through all the tutorials. What you are

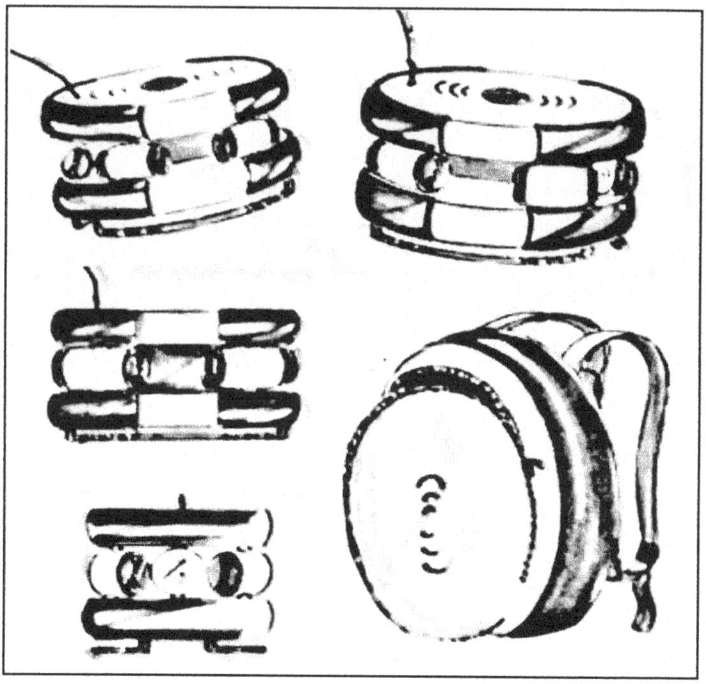

AUV Backpack Design Concept

doing is exactly what we are doing! Yes, we are also using the Arduinos and many of the same sensors you have used in the vehicle," replied Brooklyn with a smile. Turning to Cody and Derek, Brooklyn said, "I can't believe these kids are ahead of you. It is amazing what is taught in middle school these days."

Brooklyn continued the discussion. "To conclude this Zoom session. As I said earlier, I need everyone to install Fusion-360 CAD software and learn to use it. There are several excellent tutorials online to help you. Next, everyone should start investigating the Arduino boards and how to integrate the sensors. I will continue making CAD drawings of the actual vehicle.

"Diane, please send me your sketches with your backpack idea. If you want, take a picture of them and text them to me. I also need you to investigate a computer called the 'Raspberry Pi 4.' We will use it as the brain. Anthony, Melanie, and Amanda are ready to test their first software versions that they have been developing, so as soon as we can put the vehicle together, they want to test it by running it down the halls of the building, upstairs, downstairs, etc.

They programmed it like a UAV to fly through a maze without a GPS."

"Brooklyn, what did Dr. Wood say about the 3-D software?" Diane asked.

"First, he had a blank look, then laughed and said, 'Sorry, that is a no-go. I don't have that sort of money,' so we have to devise another solution. He recommended we look at the Insta360 camera, which I did. I'll send you the web link for it. It looks like a very good and inexpensive option."

ψ

Over the next two weeks, Brooklyn purchased many fundamental materials and parts. They combined and tested each sensor when they arrived. Soon, a whole table was laid out with gadgets and wires. Motors whirled, the lights went on and off, the acoustic sensors that came from their Arduino kit showed them distances, and the inertial navigation sensor showed how the vehicle was moving, such as tilting forward and back, angling left and right, rolling to either side and whether the vehicle went straight up or down.

Each time they Zoomed Brooklyn, they passed the computer camera around so everyone could see the progress. Finally, they had the basis for the vehicle.

"Now that we have it running on land, we must change it to work underwater," said Brooklyn.

Thomas quickly asked. "So, what do we need to do?"

Diane cut in. "We need to change the acoustic range finder to one that works underwater and put everything else that is not watertight into the chambers so they don't get wet."

"Excellent," replied Brooklyn. "I want you to work in groups to figure out how to waterproof each component: cameras, SONAR, LIDAR, sensors, computers, and dedicated processing boards."

"Thomas immediately cut in. "Ah, Brooklyn, what is SONAR, LIDAR, and what is a dedicated processing board?"

"That's right, sorry," said Brooklyn. "SONAR is short for sound navigation and ranging. It is when sound is used to detect an object, whereas LIDAR uses lasers, which means light imaging detection and ranging. These, along with RADAR, are the primary methods of detecting an object in air and water. A dedicated processing board is a computer that processes SONAR or LIDAR information in a form we can understand."

They all nodded their understanding, divided up the tasks, and agreed on a time to Zoom the following week.

Diane, Thomas, and Dean were to place the sonar heads that Brooklyn had made on her Fusion-360 CAD system somewhere in the vehicle.

"Guys," said Diane, "To place the sensors, we must also account for where the others should be. The interactions between sensors will be an issue."

Dean responded cautiously. "Yes, but that's going to be complicated."

"Yes." continued Diane, "but that is the only way. So, let us get started. The LIDAR mapping system is the most important, so they must be placed first. Somewhere where the system data can be combined to map the area."

After 45 minutes, they had five possible solutions.

"OK," said Diane. "Now, concerning the five LIDAR possibilities. What are the most important sensors?"

Thomas said, "Well, it depends on whether the water is very murky. If it is clear, the system will use cameras, but if it is very murky, it will switch to SONAR."

"OK, I will say the cameras since the SONAR is just a backup of the LIDAR," said Dean.

"Good point," replied Diane.

After two hours, the team came up with a couple of possibilities for each of the five LIDAR combinations.

"That's enough for today," Thomas said. "It is already past 4 p.m., and my mom expects me home. When shall we next meet?"

Dean replied, "I cannot meet tomorrow. I have a chess match, and you know I can't miss that."

"What about in two days?" Thomas asked.

"That sounds perfect. See you guys then," said Diane as she headed for the door.

She slept only a little. She was so excited. Options kept popping up in her head. The next day, in all her classes, she doodled AUV configurations with positives and negatives for each combination. When Thomas and Dean met Diane at the local Starbucks after school, Diane blurted out. "I got it guys. I have the best solution!"

"What? How so?" Thomas quickly asked.

"Here are my notes," Diane replied as she laid out page after page of notes. Some were written on napkins, and one on toilet paper. She blushed when she brought out that one as the other two laughed. The next three hours they spent going over each design until they all agreed on the best version.

As they were about to leave, Thomas said. "I was at Barnes and Noble yesterday and saw this pen that when you write in one of their special notebooks, everything you write is saved in the pen's memory. What is really cool is that you can set it to record everything said."

"Are you serious?" Dean replied.

"Not only that," continued Thomas, "it uploads everything to your smartphone or tablet when you are finished writing. I just got one and installed the app on my phone. Let's write up this last version in the notebook, then save it as a PDF file and email it to Brooklyn, Cody, and Derick. What do you think?"

"Great idea!" said Diane, "but we could use the reMarkable tablet."

"The reMarkable is very good, but what if it runs out of power? What then?" Thomas asked. "With the smartpen, it still writes since it is a normal pen with hi-tech capabilities."

ψ

Everyone had the smartpen when they Zoomed the next week, and Brooklyn had modified her CAD drawing to fit the instruments.

Cody and Derek had been working on the extra external sensors, which had also been installed on the vehicle, along with the pressure housings that would house the cameras, LIDAR, electronics, and computers.

At the end of the Zoom, Brooklyn told them all how excellent they were and that they could all take the next week off as she finalized the design. The next meeting was to be in one week, same time, same place. In the meantime, they were to investigate the software called MOOS (Mission Oriented Operating Suite), another called ROS (Robotic Operating System), and a third one called Open-JAUS, to see if this software could be used to speed up vehicle programming. Anthony, Melanie, and Amanda's vehicle control software that they had developed would be integrated into whatever they selected. Also, they would investigate software for object detection and recognition from the cameras and SONAR pictures the vehicle took. Brooklyn would look up the mapping software so the LIDAR systems could map what was before the AUV.

Everyone signed off on the latest Zoom, and Diane told Thomas and Dean: "Do you know what I haven't seen is whether the vehicle will float, be neutrally buoyant, or sink?" So, while everyone was looking things up, Diane learned how to do buoyancy and motor thrust calculations.

ψ

In the end, it was two weeks later that they were able to get together for another Zoom. They discussed all aspects of vehicle design and software implementation.

Brooklyn summarized the Zoom. "Excellent! Please send me all your files. I will start the build. Cody, please prepare all the parts that will be 3-D printed. I suggest using ABS instead of PLA. Derek, please verify all other parts that need to be machined in the machine shop. I will talk to Dr. Wood about purchasing the final materials and components, including any sensors we do not already have. Any questions?"

You will find a page describing buoyancy at the end of the book. It is easier than it looks.

Diane immediately asked: "First, what are Thomas, Dean, and I to do next? And second, what is ABS and PLA?"

"Good questions," replied Brooklyn. "Let me answer your second question first. PLA is a thermoplastic made from corn starch or sugarcane. It is a biodegradable material that is easy to print with, especially at lower temperatures than ABS. PLA parts are usually smooth and glossy, but PLA is not as strong or durable as ABS and not as heat resistant. ABS is a thermoplastic made from petroleum. It is strong and durable and is resistant to heat and impact. ABS parts usually have a matte finish and can be rough, but are stronger, more durable, and more heat resistant. If you are looking for material that is easy to print with and has good detail reproduction, then PLA is a good choice. If you need a strong, durable, and heat-resistant material, then ABS is a better choice."

Brooklyn paused for a moment and then continued. "To answer your first question. You are going to plan our sea tests. First, focus on what tasks the vehicle would do. Since all the on-land tests went well, I suggest duplicating them in a pool and then moving them to a local lake. Perhaps we can put it in the ocean in a few months."

"Neat, do we get to go too?" Diane asked. "I was in Florida last year and went to the Dry Tortugas on the Yankee Freedom."

"Of course," replied Brooklyn. "But you may have to pay your own way. I must ask Dr. Wood if he has the money to fly you all down. Florida has several underwater caves that would be perfect for testing the vehicle."

Diane, with excitement, came back, "I don't think that will be a problem; my dad would want to join in,"

"Oh, damn." came back Dean. "My dad thinks all this is just a waste of time. I must raid my savings, but I think I can make it happen."

Thomas also nodded in agreement.

The Zoom ended, and plans were made to meet the next day to set up strategies to test the vehicle.

Back in Florida, Brooklyn put together the final version of her part-list order as she finished her CAD drawings that were to be used for 3-D printing. She estimated that setting up and printing the first part for resin printing would take 18 hours.

"Hi Doug," Brooklyn began as she entered his office. Doug was one of those "Jacks-of-all-trades" people. "Can you help me with a dozen or so 3-D prints?" Brooklyn asked.

"Hmm, a dozen parts. That looks like an interesting vehicle on the cover of your document. Is that what you are going to make?" Doug asked.

Brooklyn continues with excitement. "Dr. Wood has tasked us to build a new highly maneuverable autonomous underwater vehicle with 3-D mapping capabilities using lasers and SONAR."

"So, what is the composition of the team?" Doug asked, who always wanted to ensure a project would succeed. If he found that a team lacked some specialization, such as computer programmers, he would mention it and try to help find people/students/staff/faculty with those skills.

Brooklyn continued: "Let us see, three graduate students: me, Anthony, Melanie; two undergraduates: Cody and Derek; and three middle school students: Diane, Thomas, and Dean."

Doug - If you have heard half of the stories about Doug as I have, you know that this guy is one of those rare people who are capable of amazing things. Whether diving to depths of 250 feet to some shipwreck or making an underwater archaeological survey in the Netherlands. You will find Doug among it all. His abilities are amazing. For example, we were way off the coast doing SideScan SONAR surveys. Let me step back for a moment - what is a SideScan SONAR, and what is a survey? Think of a SideScan system as a torpedo with a whole batch of sensors that project sound; the sound bounces off the bottom and is once again received by the SideScan SONAR system. Then, the sound is converted into an image, a copy of what is at the bottom of the sea, and a survey is when you cover a region. Now, let's get back to Doug. Well, the SideScan SONAR system broke. I would have cussed a lot and returned to land to send the instrument back to the manufacturer for repairs. So, what does Doug do? He gets out his screwdriver and takes it all apart. Two hours later, when I return to the lab, he is tightening the last screw as he closes it all up. Proudly stating: "OK, ready to put back in the water." Doug is truly amazing. That is when you realize that all his stories are true! *Dr. Stephen Wood - author*

"Very good; it looks like you have a good team. Now, let us look at your design," said Doug as he plugged the USB drive into the computer. "Now, what software did you use to create these files?"

"Fusion-360." Replied Brooklyn.

"OK, let me boot up the version I have on this computer to see your files," Doug said.

A few minutes later, Doug and Brooklyn were going through each file and tweaking various points for ease of part creation by the 3-D printer.

"OK Brooklyn, next, we need to convert it to a format that the 3-D printer can read. We save each file in a special format and run it through another software program to verify that the 3-D printer can print it."

Two hours later, they had finished the review and conversion. Now, they were ready to print the first part.

Hours later, out came the first part. "Now Brooklyn, this was a wide mesh, some may say low infill, ABS print where we will fill the wide openings with epoxy," said Doug.

Doug continued, "We could make resin prints of each, but it would take more time, and my printer cannot print pieces as big as the one we just printed."

"OK, Doug," Brooklyn replied.

Doug continued, "Let's print them all with a wide mesh and fill them with epoxy. If they work, we can use them to create a mold that would allow us to easily make many of them by pouring plastic into them. A molded plastic part is straightforward to make."

Two days later, she worked on finalizing the parts that would require a machine at the machine shop.

As the materials arrived in the mail, she tasked Cody and Derek to machine various parts at the machine shop on the other side of campus. They carefully machined each piece using the CNC (Computer Numerical Control) mill, hand-controlled mill, and lathe.

As all the parts came together, Brooklyn met daily with Amanda, Anthony, and Melanie to integrate the vehicle navigation and control software and the many sensors. The team decided to use a combination of MOOS and ROS with integrated navigation and control algorithms from the three graduate students. Soon, the vehicle looked like an AUV with a working brain. After a week of talking to each other about test possibilities, Diane, Thomas, and Dean came up with an easily deployed grid system for testing in the pool that will work with lasers. The sides would be made of fabric held up by plastic tubes. When they want to test the acoustics system, the sound waves will go through the fabric, so the fabric would have to be replaced by some hard material like thin sheets of hard plastic or aluminum.

Hi guys!" Brooklyn said one day at the end of June. "Congratulations! The assembly and programming are complete. You all did a fantastic job. I talked to Dr. Wood this morning. He has the money to fund the three of you to fly to Florida to participate in the vehicle tests. Are you interested? I know you are, but he must have written documentation from you and your parents consenting to the trip. I will email you all the forms you and your parents will sign. Send them all by noon tomorrow, and we will have flights arranged for early next month, sometime around the 6th of July. Expect to stay for about four weeks. Everyone in?"

Emphatic yeses were heard all around.

"Excellent! You guys are great. I will send you the information as soon as we sign off. Any questions?" Brooklyn asked.

Ten minutes later, Brooklyn was emailing the documents.

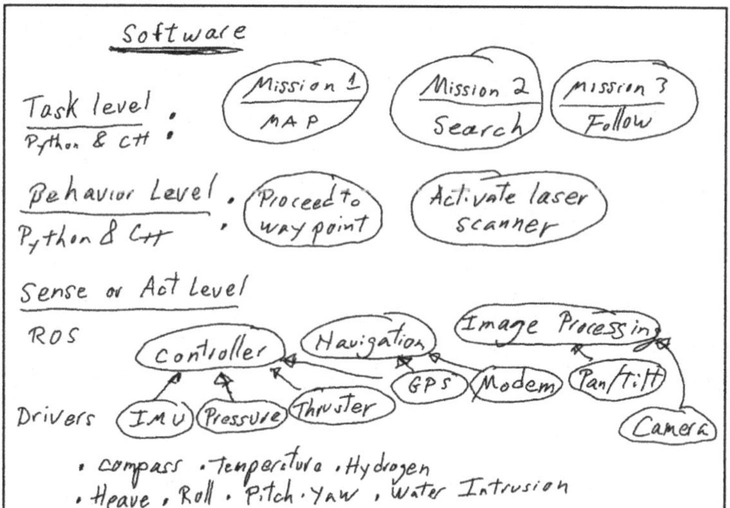

Software Modules

INTO THE DEPTHS

It was their last Zoom meeting before flying to Florida.

"Hello everyone," Brooklyn said.

"Hi," responded Diane, Thomas, and Dean.

"Great to see you all," continued Brooklyn. "You have your electronic plane tickets. Housing has been provided at the university, along with food. As you saw, I emailed you a list of things you will need—primarily multiple bathing suits. I will have plenty of sunblock, up to SPF 100. Bring a couple of pairs of sunglasses with restraining straps for them and, of course, two hats."

"Why do we need more than one hat?" Dean asked.

"Because you will lose at least one of them, so don't bring your favorite ones," replied Brooklyn. "Expect to get dirty wet and have a fantastic time."

ψ

9 a.m. Sunday, July 6th

Sam, Suzanne, Diane's brother Bud, and their dog Freddy jumped into their minivan with Diane and drove to pick up Thomas and Dean, where their parents said their goodbyes, gave them big hugs, and told them to call or Zoom every day.

On arrival at the airport, they made their way to security. Suzanne gave Diane the biggest hug and gave Diane some parting advice to be careful.

Sam went through security with the kids and went to the gate.

"Diane," said Sam. "I know you can handle yourself, but be careful. All of you stick together like glue. You will all have some amazing experiences. Please keep us updated on what you are doing. We want to hear from you often. Text us and send pictures. If you need anything, holler. If you need help, I can be there in a couple of hours. Diane, here is a credit card if you need to buy things. Since you are all underage, always keep your permission letters with you. It would be best to take your passports with you for identification. Please keep all your documents and money in the waterproof pouches we gave you. Keep all the receipts you receive for everything you buy. All of them. As I said before, if you need anything, holler. Now, Diane, hug your dad, and off you go. It looks like they are boarding the plane."

"Love you, Dad," said Diane. "We will text you when we land and meet Dr. Wood and Brooklyn." Diane, Thomas, and Dean got in line and soon were out of sight.

Sam turned away with a tear in his eye. "That's my girl!" He said to himself as he returned to his wife Suzanne, their son, and their dog Freddy.

On the plane, one of the flight attendants led them to their seats. There, they settled in for the three-hour flight. It was Thomas' first flight. Diane and Dean filled him in on all the details, showed him the emergency and in-flight magazine, and continued their discussion about the project. It was their first trip without their parents.

"Thomas, it's so wonderful that you had the project where you got to know Dr. Wood," said Diane.

"Well Diane, it is also great that you have a father who likes to invent things. I'm sure that when Dr. Wood and your dad find time to meet, all kinds of cool things will come out of it," said Dean.

Thomas continued. "Yeah, Diane, your knowledge of electronics and computers that you acquired while working with your dad is impressive."

The last part of the flight was rather bouncy. The afternoon thunderstorms had built up around Orlando like they always did during the summer. Taxiing took a long time since they had landed on the far eastern runway, so they had plenty of time to text their parents that they had arrived.

On leaving the plane, they jumped on the tram. Thomas was in the front of the tram, pointing out everything, especially the palm trees.

As they passed security and left the boarding area, they saw Brooklyn and Dr. Wood waiting.

"Hi everyone!" said Dr. Wood.

“Hi’s” were said all around.

“How was your flight?” Dr. Wood asked.

“Bumpy,” Dean responded, who was closest to Dr. Wood.

"Well, you didn't throw up, eh," replied Dr. Wood. “So, it could not have been so bad. I remember a flight when I returned from an adventure in the Ukraine. Everyone around me was barfing. The guy next to me was white as a sheet.”

“Ukraine! What were you doing there? Are they not at war with Russia?” Dean asked.

“Well, there was this submarine that I wanted to look at in Odessa. Wonderful people. Someday, I may go back if the war ever stops,” replied Dr. Wood and a minute later asked, “By the way Diane, where is Bronwyn?”

“Bronwyn? How do you know Bronwyn?”

“We all met in the Florida Keys last winter while you were on spring break. If I recall correctly, you were researching global warming, coral reefs, and sea level rise.”

“I don’t remember. . .”

“Your boat had to tie off on the research vessel Hogarth. I was the principal investigator on the ship,” said Dr. Wood.

“That was you! Yes, I remember now,” said Diane, beaming. “To answer your question, Bronwyn would be here, but her parents took her to some place in Idaho. In the middle of nowhere. Something about contract work that paid big money.”

“Oh, too bad. You worked very well as a team.”

They picked up their luggage, jumped into Dr. Wood’s Subaru Outback, and headed out into a thunder and lightning storm.

While Dr. Wood drove, Brooklyn outlined the following week’s activities: first, housing; second, food; third, meeting locations and times; vehicle completion, tests, and what challenges they still have. She had so much to cover that no one realized how fast time had passed. An hour and fifteen minutes later, Dr. Wood drove into the Underwater Technology parking lot of the university.

"Brooklyn, here are the keys to the dormitory room. The boys will be roomed together, and Diane will be across the hall. I will let you take care of the rest," said Dr. Wood.

"See you all in the morning, say 9 a.m., in the dining hall. Bye."

Dr. Wood returned to his car.

ψ

9 a.m. Monday, July 7th – They met at the dining hall for breakfast.

"All right guys, eat as much as you like. We have a couple of long days ahead of us," said Brooklyn. "Dr. Wood texted me. He is off to La Spezia, Italy, to talk to an engineer at NATO about issues with our docking station for our REMUS 100 autonomous underwater vehicle. So, he will not be here to hold your hands or wipe your noses."

Brooklyn handed out a couple of pages with times and places on them and spiral-bound booklets as two other students came to the table.

"Hi everyone," said Cody and Derek.

"Finally, you guys got here," said Brooklyn. "Here are your packets."

"Hi Brooklyn, when is our first break? Derek and I want to hit the pool sometime this morning," said Cody.

"Check your schedules I just gave you. We are all going to the pool at 3 p.m. Lunch is at noon," said Brooklyn.

"OK, but that is getting closer to when the afternoon thunderstorms are supposed to arrive," said Cody.

Over breakfast, they outlined the morning activities. When their breakfast was over, Cody and Derek headed to the machine shop while Brooklyn took Diane, Dean, and Thomas to the laboratory where the 3-D printers could be found.

Brooklyn sat with Diane, Thomas, and Dean at the Underwater Technology Lab. The table had four laptop computers on it. Each had a sticker on it.

"I love the stickers you put on these laptops! Pinky and the Brain are riding a yellow torpedo! It looks like Pinky is holding on for dear life, and the Brain is plotting to take over the world," said Diane.

"Actually, the yellow torpedo is an autonomous underwater vehicle made by Bluefin, a company based in Boston, Massachusetts. We had two of them," said Brooklyn.

"Had?" Thomas asked.

"Well, we now have only one. The original owners put the first Bluefin in the water without properly closing the pressure chamber. We are putting it back together. Hopefully, we will have it running by this December," said Brooklyn.

"Are you serious?" Thomas asked.

"Yes." Continued Brooklyn. "Two of Dr. Wood's graduate students decided to modify the navigational program of the vehicle. Well, they hacked into the system after guessing the master password. They modified the code and took it out to sea a day later to see how their modifications would work. Well, it ran the first 100 yards okay, turned and dove, and was never seen again. Of course, the graduate students did not bring the emergency abort SONAR system that would allow them to force the vehicle to drop its heavy ballast, which would force the vehicle to the surface."

"Oh, crap. Dr. Wood must have been furious!" Thomas said.

"Well, to say the least, he was not pleased, but as he always says, 'If you don't want to lose it, don't put it in the water,'" said Brooklyn.

Dean promptly asked. "Brooklyn, is that the other Bluefin AUV over there?"

"Yes," said Brooklyn. "Dr. Wood has two other graduate students assembling the electronics to get it running again. Just like in our system, they will use Arduino boards to run the individual components and a Raspberry Pi running Linux as the main onboard computer."

"What! They are using a $60 computer to run a million-dollar vehicle!" Diane said.

"Yup. Many aspects of our system are directly transferable to their system," replied Brooklyn. "For example, we use the same IMU."

"An IMU, what is that?" Thomas asked.

"I can answer that," said Diane. "I read about them while looking up the basics of AUVs. It is called an Inertial Measurement Unit that is part of your navigation system."

"Quite right, but there is more to it than that, but we don't have time to go into the details. This is the one we use. A Sparton AHRS-8 that costs around $2,000 each," said Brooklyn. "Here is the one we will use."

"What is this little thing?" Dean asked.

"Yes, this little thing," mocked Brooklyn. "And this little thing here is the pressure or depth gage, and this little thing measures if there is any hydrogen building up in the battery chambers, and this little thing detects if we have any water in the pressure housing where all the electronics are, and these three things are called hydrophones and with proper programming can determine if we hear Billy Bob's fishing boat cruising by or whether it is Peggy Sue's shrimp boat or something else such as an enemy submarine entering our harbor. The hydrophones listen to the frequency of the boat's propeller blade and can recognize differences in their sounds."

"Really?" Diane asked. "How is that done?"

"By analyzing the signals using something called neural networks. Advanced stuff," said Brooklyn. "But let us get back to the here and now. These laptop computers are yours to use while working on the project. They have everything you need, such as Fusion-360 solid modeling computer-aided design system, C++, Python, MATLAB, LABVIEW, the Arduino user interface, WORD, POWERPOINT, EXCEL, and EMAIL. It has one Terabyte of cloud storage. The computer is the latest you can get, with the fastest processor, with enough RAM (memory for the computer) and storage for anything you plan to do. Here is its case: mouse, mouse pad, extra hard drive, and a USB thumb drive. Keep them all safe."

"Wow," said Diane. "We never expected this."

"Of course not, but that is how Dr. Wood is. He wants us to succeed and gives us the tools to make it happen. For example, see these pink and gecko-green toolboxes. They are filled with every tool you will ever need. Keep them well and safe," said Brooklyn.

"Now everyone. Come to this table. Here are most of the parts for the vehicle. Well, most of them. You should boot up your computers and go to the CAVE-AUV document. It has all the current documents, such as CAD drawings and parts. Let me know if I have missed anything. Be very thorough," said Brooklyn.

They powered up the computers, and Brooklyn showed them where to find the files. An hour later, they finished. Only a few minor items were missing. Diane pointed out that the list contained only some of the computers that would go into the pressure housings.

"Very good," said Brooklyn. "I will have to update the list and will get you the updated one tomorrow. Right now, I want you to start assembling the vehicle."

Diane started by placing the multiple 3-D printed parts in a group. Dean lined up the thrusters, batteries, and external components and sensors. Thomas took control of the pressure chambers and internal sensors and components. Five hours later, all the components were in their locations.

"All right, the pieces that didn't fit easily, I need them listed for reworking," Brooklyn said. "Tomorrow, we will work on the electronics, but now is the time to head to the beach. Everyone, get your swimsuit. Let's meet back here in twenty minutes."

HOT CHICKS - COOL DUDES

4 p.m. Monday, July 7th

After a late afternoon of swimming and frolicking in the ocean waves, Brooklyn, Cody, and Derek pulled out the barbecue coals and started a fire for an evening dinner of burgers, hot dogs, and various fixings. It was late, around 11 p.m., when Cody said, "It is time to walk the beach, guys. There will be a full moon tonight, so we should have good visibility."

"For what?" Diane asked.

"To see sea turtles come up and lay their eggs."

"Are you kidding me?" Dean asked.

"I always wanted to see a sea turtle," said Diane.

"The chances are high that we will see one. We will probably see more than one," said Brooklyn. "I have been a Sea Turtle Preservation Society member for five years. I have just been promoted to be able to show the mothers laying their eggs. Since I have been added to the permit, I will assist Richard, who is the primary permit holder, in showing a turtle tonight. Richard is the best! He was the one who suggested we show the sea turtle egg-laying experience."

"You are serious!" Thomas exclaimed.

"Yes, very serious. And here comes Richard," said Brooklyn.

Richard parked his car and came over with a box of stuff.

"Hi everyone; nice to meet you all."

Brooklyn made the introductions, and then Richard took to the floor.

"Excellent. First, we will search for the *Caretta Caretta* sea turtle, also known as the loggerhead sea turtle. They are often more than a meter long, or the shell is, and weigh around 350 pounds."

"Wow," said Dean.

Richard continued. "The loggerhead is the most common and is only considered threatened in this area. The other sea turtles are all classified as endangered. These are the Green Turtle, the huge Leatherback, the Hawksbill, and the rare Kemp's Ridley."

"How do we know what type we will find?" Diane asked.

"Well," said Richard. "It's the wrong season for the Leatherback. They usually appear in the winter or late spring. Their nests are huge. Well, not the nest but the area they disturb when they lay their eggs. All sea turtles try to disguise the area where they lay their eggs so that no one or nothing can find them."

"So, how will we find the eggs if they are camouflaged?"

"Tonight, I will show you once we find our 'little lady,'" said Richard. "It is unlikely that we will see a Green since they usually come up late in summer and fall, but it is possible. Regarding the Hawk's Bill and the Kemp's Ridley, they are extremely rare and much smaller than the others."

Diane jumped in. "So, what do we do tonight?"

Richard looked at Diane and said: "Well, we first walk the beach. Brooklyn, Thomas, and Derek will go north, and Diane, Cody, and Dean will accompany me to the south. We will walk along the water's edge until we see a turtle or the tracks of a turtle. Brooklyn has one walkie-talkie and me the other. When we find one, the other group comes and joins. During this time, either Brooklyn or I will check out the tracks. If they look like a huge truck tire, it is a Green, and we will skip it. If the individual tracks look like a crescent moon, then we know it is a loggerhead, and off we go."

"So, what is in your pouch that you brought?"

"Walkie-talkies, red flashlights, and a medical kit in case one of you step on something sharp. Okay everyone, to the beach."

"Quick question," Diane said. "Why red?"

"Quick answer. Because scientists have shown that sea turtles supposedly can't see red," said Richard.

The two teams headed off, one south and one north. It took only 15 minutes when Richard pointed to a lump at the water's edge.

"There she is. Just coming out of the water."

Sea Turtle Tracks

“You mean that lump way off there? I can barely see it. Is that a sea turtle?” Cody asked.

“Yes. Diane, here is the walkie-talkie. Call Brooklyn and tell her that we found one and that her team should come here.”

“Sure,” answered Diane with some puzzlement. “How does the Walkie-Talkie work? I have never used one before.”

“Easy. See this switch on the top. Rotate it clockwise. It turns on, and the more you turn, the more it increases the volume. Now, on the left side, you will see a large button. Push it and hold it down. Put the walkie-talkie close to your mouth and speak normally. Start with something like ‘North group, North group; this is South group, do you hear us?’ Then, let up on the button and wait. They should respond with, ‘This is the North group; we hear you loud and clear.’ Then hold the button down again and say your message: 'South group. This is the North group; we have a turtle. You will find us after a 15-minute walk north of where we started.' All right, here is the walkie-talkie.”

Diane took the walkie-talkie and spoke exactly as Richard had instructed, and soon, the north group was headed south. Richard, Diane, Cody, and Dean watched as the turtle slowly came out of the water, looking around.

Richard said. "I remember a few years ago that I was with Dr. Wood when I was training him."

"What! You trained Dr. Wood? He was in the Sea Turtle Preservation Society?" Diane asked.

"Oh yes. He was in for over ten years. So, there we were; we had searched and searched for a turtle. Finally, around midnight, up came this beautiful little lady. There were several people on the beach even though it was so late. We were watching the turtle, but not everyone else. Suddenly, a runner zoomed by us and ran right in front of her nose. Normally, such an event would cause the turtle to turn around and head out to sea, but no, she stays. She just pauses for a few minutes and is just about to continue when, once again, the runner zooms back right in front of her. Crap! What an idiot! We waited and waited, keeping our fingers crossed that she would stay, and yes, she did. After about five minutes, she takes about four steps when, zoom, a bicycle passes by with those huge tires they have for riding on the beach - they should be banned! The bicyclist swerves so he wouldn't run over her. Right in front of her nose. Just a few inches away. We curse that rider with many foul words. But the turtle stays! Unbelievable! The bicyclist turns around about two hundred yards down and zooms again past the sea turtle. We wanted to punch this guy, but he was able to swerve away from us. We think that this turtle is not going to stay. Any one of these events would normally cause the turtle to go back out to sea. But we have amazing luck! She stays! So now it is very late. The people are gone, and she heads up the beach. Then, just as she makes it to the high tide line, we see off in the distance two bright lights; a minute later, two Apache helicopters zoom by just a hundred feet up, loud as a freight train zooming by with two super bright spotlights shining down. That was it! She must have thought that enough is enough. Two runners, two cyclists, and now two helicopters; I am out of here, so she returned to the sea, and we went home."

"OMG!" said Diane. "That is unbelievable! What callous people. Mr. Richard, what happens to the sea turtle that is scared and goes back out to sea?" Diane asked.

"Just call me Richard. The sea turtle will return someplace on the beach later that night or perhaps the next night. Still, I have been diving off the coast here during the turtle season and have found hundreds of dead eggs in the water where she finally dumped them because she was constantly threatened when she went ashore!"

“That is awful!” Diane exclaimed.

The north group came up to the south group at that moment.

“Where is she, and what is she?” Brooklyn asked.

“She is almost at the dune line. It is great that we have almost a full moon. We can see everything. Brooklyn, take Diane along the water line until you reach her tracks. Then come back and tell us if she is really a loggerhead turtle!” Richard said.

Diane and Brooklyn made their way along the water's edge to her tracks coming out of the ocean.

Brooklyn asked Diane. “So, what do you see? What is special about these tracks?”

Diane studied the tracks intensely.

Brooklyn asked again. “What is so special about those tracks?”

“Well,” began Diane. “First, each flipper print has a curve.”

“Excellent. What does that mean?” Brooklyn asked.

“That it is a loggerhead?” Diane responded more as a question than an answer,

“Yes, exactly. Now, what else do you see?”

Diane continued. “She must be really big considering the spread of her flippers and... now this is weird. Her right rear flipper must be hurt.”

“Why would you say that?” Brooklyn asked.

“Well, look at this mark. It looks something like what a wooden peg leg would make,” replied Diane, pointing to the impression in the sand.

“Excellent! That is correct. We saw her two weeks ago when she came up and laid her eggs.”

“Then why is she here?” Diane asked.

“To lay more eggs. See, some female sea turtles lay up to six nests in a season.”

“Wow! And how many eggs?”

“Somewhere around a hundred. Sometimes less, sometimes more,” answered Brooklyn. “Let's go back and report what we found.”

They waited with the others until the Loggerhead turtle found her spot near the dune line.

"Ok Brooklyn, this is your night. Take Diane with you as a messenger," said Richard.

"Ok," replied Brooklyn. "Diane, follow me, do what I do, and do not talk unless I ask you a question."

Brooklyn and Diane made their way to where they had analyzed the tracks. Brooklyn gave Diane some instructions. "Diane, you are the messenger. I will tell you what you will need to relay to Richard. We will be approaching the turtle and need to communicate with the others. You are the messenger, so come with me to the turtle. From the water's edge to the soft sand line, we crouch. From then on, we must go commando on our bellies until we can see what the turtle is doing."

"What do you mean by 'doing'?" Diane asked.

"Well, she doesn't just go up and lay her eggs. First, she checks out the area. If it looks good, she will swivel left and right, flattening the area a bit. Next, she will start digging. You know this because she will start throwing sand. First, one hind flipper, and then the other. She will take her flipper and scoop the sand as you would by cupping her flipper. The hole she makes is about a foot across. As she gets deeper, her shell, well, her body will angle higher as she reaches deep in the hole. That is when you know she is getting close to laying her eggs. At some point, she will stop digging. She will then rest for a moment. At that time, we want to slide up to her rear and check out her rear flippers."

Brooklyn paused to let this sink in and continued, "When you see both flippers rise slightly simultaneously, you know she just dropped some eggs. At that time, you will go and get the others. Have them come up next to the tracks. Please do not allow them to disturb the tracks. The surveyors who will come by to count the nest early tomorrow morning will need to see them. While you are doing that, I will prepare the nest for viewing. I will remove some sand just behind her, but not enough to disturb her. I will smooth it out and place this small flashlight with a red filter to shine on the eggs. When you arrive, I want you to sit beside the right-hand rear flipper. Do not let anyone pass you. We want to disturb the turtle as little as possible. Sit there quietly until she is done. She often drops two or three eggs at a time until the nest is three-quarters full. Understand?" Brooklyn asked.

Diane nodded, and Brooklyn continued, "When she is finished, she will stop briefly and then fill the hole. Quickly grab the flashlight before she sweeps it into the hole. It will take about five minutes to fill the hole. Next, she will take both rear flippers and pack down the sand using her whole body, raise it, and slam it down to further pack the sand. She will do this a few times. She will then rest for a minute or two. Then, she will start flipping sand all over the place with her front flippers, throwing sand over her back and hind area as she slowly moves forward until she is around ten feet from the nest. The entire area will be covered with sand. Finally, she will stop and rest for a moment and then head to the open ocean. She then slips under the ocean waves until the next time."

Diane said, "OK," with a great smile.

They made their way up the beach. Brooklyn and Diane quickly made their way to the high tide line, then kneeling, they made their way for a few more feet. Because of the moonshine glistening off the turtle's shell, they knew precisely where she was. Now, going on their bellies, they made it close to the hind end of the turtle. She was already tossing sand, first left and then nothing right. Then left and nothing right. Repeatedly. Finally, after what seemed to be an hour, she finally stopped.

Brooklyn and Diane silently crept forward inch by inch until they were within arm's length of the turtle. Within a minute, they watched the rear left flipper curve slightly upward.

Brooklyn whispered to Diane. "It has started. She has just dropped her first set of eggs. Go at a crouch and bring the others here. I will prepare the site for viewing."

Diane returned to the water's edge, where she found the others waiting.

"How is it looking?" Richard asked.

"Brooklyn said that she had just dropped her first eggs," said Diane.

"Excellent," said Richard. "Everyone, follow Diane up the tracks. Do not walk on the tracks. A team will be going up the beach at dawn tomorrow to survey how many nests there are, and we don't want them to miss this one. And, as I said earlier, keep your voices to a whisper, no lights, no flash photos, and do not go beyond Diane and Brooklyn."

Diane followed the tracks until they reached Brooklyn and the sea turtle. Diane sat inches from one rear flipper and Brooklyn from the other flipper. A red light shone down the slope to the turtle's hole.

Suddenly, three white ping-pong ball-sized eggs dropped down the hole. Words like "Wow!" "Amazing," "I always wanted to see this," and "Just like on TV" was whispered. The eggs, after about twenty minutes, were piling up. You could see that there were close to 100, if not more.

One of the kids watching the eggs drop asked Richard, "So how can you tell which eggs will be girls and which will be boys?"

"Well, it comes down to how warm or cool the nest is. If the nest is warm, there are usually more female babies. If the nest is cool, then there will be more male babies. This is why we have the term 'Hot Chicks, Cool Dudes.'

Richard quietly said: "She is almost done. Brooklyn, could you take out the measuring tape?"

Brooklyn pulled out a tape from her pocket. On the one side, there were inches; on the other, there were centimeters.

"Brooklyn, which do we use? Centimeters or inches?" Richard asked.

"Centimeters, of course," answered Brooklyn.

"Quite right. It is time to measure her."

Brooklyn stretched the measuring tape from top to tip lengthways up her back.

"Anyone got a light?" Brooklyn asked.

Richard came over and briefly flashed his light. "One hundred and fifteen centimeters. She is a biggie," Richard said.

Brooklyn sat down again just as the turtle took a swipe with her right rear flipper. Nothing happened.

Richard quickly said. "Quick, get the light. She is about to cover the nest."

Diane grabbed it a second before the left flipper came across, filling everything with sand.

Brooklyn asked Diane, "Diane, help me push back the sand I removed to show the nest. Working together, they made it look as if nothing was disturbed.

Five to ten minutes later, the hole was full. They watched as her entire shell lifted as she packed down the sand. Soon, she was flipping sand everywhere as she slowly moved ten feet forward. Finally, she paused, looked around, and returned to the ocean at top turtle speed. As she slipped into the water, Richard had Diane light up the rear half of the turtle with the red-filtered flashlight. Into the water, she went. Then, she was gone.

"All right everyone, back to the cars. It is time to go home," said Brooklyn.

"Keep to the waterline and keep your eyes open for sea turtles coming out of the ocean as we head back."

Fifteen minutes later, they washed their feet off and climbed into their cars. Brooklyn and Diane had sand everywhere: in their hair, down their shirts, pants, everywhere.

DANCE THE NIGHT AWAY

11 a.m. Tuesday, July 8th

Everyone met in the dining hall at the early hour of 11 a.m. the following day. They were still excited about their experiences when they returned to their dorms and talked until 3 a.m.

"OK, everyone," started Brooklyn.

"Cody, were you able to finish the parts you were machining."

"No, but we will be done by 5 p.m. today," replied Cody.

"Good. Bring everything to the lab tomorrow morning. I also have a few parts that need slight modifications. Derek, come by the lab when you have a break to pick them up. Detailed descriptions are written of what needs to be modified. I have already updated the CAD files."

"What are we doing today?" Diane asked.

"Electronics with Doug. Doug is our electronics specialist. We will spend today and tomorrow testing and connecting each sensor and electrical component, starting with the motors. Then, the inertial navigation component, SONAR transducers, lasers, etc. We have already tested how each works and have implemented software routines for the vehicle to access and control each. The question is to combine them one by one, connecting each and then verify them in combination."

"Isn't it faster if you just connect all of them and turn them on?" Dean asked.

"Definitely not! In fact, from my experience, it will take a lot longer to find out where the problems are," replied Brooklyn.

Diane finished her mushroom omelet and the hashbrowns on her plate, drank her orange juice, and joined the others in returning all their plates and utensils.

Back at the lab, they met Doug, who was already setting up the oscilloscope, frequency generator, power supply, and soldering iron.

"OK, let's start with the internal components of the main pressure housing: Raspberry Pi 4 and Arduino boards that will control the motors," said Doug.

An hour later, they were satisfied that the motors were running fine and could be controlled remotely if desired. Otherwise, the Raspberry Pi would control them.

"What next?"

"Good question," Doug answered. "I would say Spartan's IMU. This way, movement and navigation sensors would be implemented. Then we should include the internal sensors that will go into the pressure housing: water intrusion detection, hydrogen level, and temperature."

By 5 p.m., all internal sensors were integrated.

"I am so glad we spent the time understanding each of the sensors before we tried to integrate them," said Brooklyn.

"Yes, testing and understanding each component before integration are essential. Then, you can write the programming code to control the sensor or use its data. Those hours you spent with each component or sensor were well worth it."

"Shall we start with the external components?" Brooklyn asked.

"No, let's end on a high. I will see you all tomorrow at 9 a.m.," said Doug.

Suddenly, they heard a boom. Then, with a flash and another boom, followed immediately by two more, the rain pelted the windows as gust after gust blew the rain against the lab window.

"Wow!" Diane exclaimed. "It was sunny a moment ago."

"Ah, yes; a typical Florida summer afternoon," said Brooklyn. "It's lucky we just finished. All we need would be an electrical power spike from a lightning bolt hitting a power line, and all the work we did today would be gone, and we would probably have to order more electronic parts since they would all be burned out."

"What should we do now?" Dean asked.

"Sit back and enjoy the show. There is not much else we can do unless you want to get really wet," said Doug. So, they sat back and watched the lightning and rain pouring down. Half an hour later, it was over, and blue skies were seen through the clouds. The air was cool and fresh as they returned to the dining hall.

ψ

9 a.m. Wednesday, July 9th – They returned to the lab again. Doug, Brooklyn, and Diane worked on the laser system while Cody, Derek, Dean, and Thomas assembled the entire vehicle. Placing the SONAR transducers in their spots, mounting the video cameras and the extremely close proximity feelers, the CTD, that is, conductivity, temperature, and depth sensor that will document the speed of sound at its location; and a DVL, or Doppler Velocity Logger, to measure the vehicle's velocity. With only a short break for lunch, they worked steadily until 4 p.m. when Doug said: "Okay guys, that is enough. Off to the pool, all of you! See you all tomorrow morning at the same time. We should be able to finish the vehicle's electronics and mechanics by the day after tomorrow. Then, we will work through the programming. Anthony, Melanie, and Amanda put together a great navigation and control algorithm; now we need to verify and test it first in the lab and then in the field."

Everyone ran to their dorms and met at the pool in 15 minutes. Brooklyn brought a ball and two floating goals. "Okay, everyone. We are playing modified water polo. As you can see, this pool section is only four feet deep. You can move if you do not have the ball. You can only move three steps with the ball."

Cody said. "And no roughhousing. Dr. Wood will be extremely upset if he has to go to the hospital or the police station."

Diane immediately said. "I am the captain of team 1."

Brooklyn said. "I call dibs on team 2 captain."

Team 1 consisted of Diane, Cody, and Dean. Team two consisted of Thomas, Derek, and Brooklyn.

Who won the game was anyone's guess since it was halted in a fit of laughter when, after a rough session of trying to get the ball, Derek's swim shorts were torn so severely he had to hold them together with both hands.

As they walked back to the dorm, Diane asked Brooklyn. "Brooklyn, I couldn't help noticing a flash of silver when Derek lost his shorts. What did you do?"

Brooklyn just gave Diane a wink and changed the conversation.

ψ

9 a.m. Saturday, July 12th – Three days later, they had all the AUV components working together. They worked from 9 a.m. to midnight for the last two days, with breaks just for lunch and dinner. Problems popped up everywhere. As soon as one was solved, another was immediately there.

With all his years of experience, Doug saved the day by realizing that a whole package of newly purchased electronic diodes was defective. Luckily, Brooklyn found some in Dr. Wood's supply in the lab. It was late Saturday afternoon when Doug and Brooklyn gave each other a high-five.'

"Yes, yes, it works! Damn, that was difficult to find. Thanks for all your help, and thank you all for being so patient. Ok, we all need a break. Brooklyn and Cody, why don't you take Diane, Thomas, and Dean to Kennedy Space Center tomorrow and make a day of it," said Doug.

"Kennedy Space Center! Yes, that is a great idea, and from how you are all jumping up and down with excitement, I guess you have no objections," said Brooklyn.

"Brooklyn, after we finish dinner, we want to go to the movies," said Diane.

"I can't make it," said Brooklyn. "I have dinner plans tonight. Cody, can you drive them?"

"Sorry guys. This is the night I will be in Orlando dancing," Cody replied.

"Dancing, that sounds great. Can we come?" Diane asked.

"Do you know how to swing, foxtrot, or waltz? Tango? What about Cha Cha, Rhumba, Quickstep, or Paso Doble? That is what I will be doing," replied Cody.

"Count us out," said Thomas and Dean.

"Well," said Diane. "My mom and dad taught me some moves in each of them. I am not very good, but I know the basics. Can I come?"

"OK," replied Cody. Get changed into something nice and wear good shoes, preferably with leather or suede bottoms. Return in an hour. Don't be late. There are several ladies who want to get plenty of dances in before the night is over."

Diane ran out of the lab to her dorm room. Derek agreed to drive, and Thomas and Dean went to the movies.

ψ

7 p.m. Saturday, July 12th – One hour later, Diane and Cody were in Cody's SUV on the way to Orlando for ballroom dancing.

Cody arrived on time to find Diane looking lovely in a light blue dress with matching shoes.

"Perfect, you look great," said Cody. "You are going to be noticed by a lot of guys. Just tell them you are a beginner, and by the end of tonight, you will be an intermediate dancer."

"Cody, how did you get into ballroom dancing? No one ballroom dances." Diane asked.

"Actually, a good number of young people do. As you will soon see. Especially those who do swing dancing. Some are really, really good," said Cody.

"But how did you get into it?" she asked again.

"Well, a couple of years ago, I went dancing. I asked this girl to dance. She said yes, but she was looking everywhere but at me the entire time. It got so bad that I just walked off the dance floor, and she didn't notice for a few minutes. Then, a week later, I asked two women to dance with me. I thought I would surely get in an evening of dances. But no, two guys knew how to swing dance. I had one dance with each of them, and then they danced with the two dancers for the rest of the night. That is when I said to myself: 'I must learn that!' A month or two later, my gym had an ad saying that if you renewed your membership this year, you could take their new ballroom dancing class for free. The rest is history."

"That was not a nice person," said Diane.

“I thought so too, but it is often typical,” replied Cody with disgust and frustration.

An hour later, they arrived at their destination. They got in line to pay the entrance fee and looked for a table. Two women bounded over to them with great smiles on their faces.

“Cody, you made it! We heard you might be tied up this weekend,” Gwen and Sabrina said excitedly.

“Nope, I got out of it, but let me introduce you to Diane. She is on the design team. She knows some ballroom moves. Perhaps you can get Lance to dance with her,” said Cody.

“Hi Diane. Have you ever been to a ballroom dance before?” Sabrina asked.

“No, but my mom and dad did. When they got home, they showed me all types of moves. When I was little, my dad tossed me all over the place. He always used to shoot me through his legs. Once, he had me do a handstand on his shoulders. I was so high up! Oh my God, my mom was furious! It was weeks before my mom allowed him to dance with me again,” said Diane.

Sabrina looked at Gwen. “Did you hear what I heard? She has already done some very advanced stuff and doesn't even know it!”

“Cody, get your shoes on. Sabrina and I will be back in 10 minutes. Diane, come with us,” said Gwen.

Gwen, Sabrina, and Diane headed to the far side, where a group of teenagers were.

“Lance, let me introduce you to Diane. She knows some ballroom moves, but we think she actually knows a lot of swing but does not really know it. We don't think you will complain,” said Gwen.

“Diane, this is Lance. He is one of the finest dancers here and probably the most radical. Also,” Sabrina whispered to Diane, “he is very handsome.”

“Come back to our table when the mixer starts,” said Gwen.

Gwen and Sabrina gracefully glided back to Cody, who took them both onto the dance floor for an Argentine tango, with the women interchanging with each other while Diane and Lance watched.

Lance commented. “Oh, that Cody can really dance. You will not find many like him. He is pretty unconventional. Like that move there, he made that up, but boy, wasn't that smooth! Diane, look at the subtle details of his movements and how he controls the dance.

"Sabrina and Gwen are also fantastic. Otherwise, the dance would be a disaster. Look at those leg flips. And now, look at Sabrina sliding her leg up Cody's. Now, what you just saw was fantastic dancing."

Cody, Sabine, and Gwen left the dance floor. Next, a Waltz was played, and each reappeared on the dance floor with new partners.

"OK, Diane, let's dance," said Lance.

And the two went to the center of the dance floor.

"Can't we dance out of the way on the side?" Diane asked.

Lance responded with a smile. "The center is out of the way for the Waltz. You see, the flow of dance, for dances that move, is counterclockwise. The slower you are, the closer to the middle you should be."

"Oh," said Diane.

Lance took Diane's right hand in his left and put his right hand on her back just where it met the shoulder blade.

"Now, Diane put your left hand on my right arm. Rest it lightly there, very lightly. You will use that connection to determine some of what my actions will tell you. Your left hand connects with my right, like so. Now, lean back slightly and look over my right shoulder. Remember that I will be leading. Relax and feel the beat. One-two-three, one-two-three. Feel the motion and the transitions I will have you make. Be sure to hold your frame. Like this."

Lance swayed left and right with a firm, professional frame.

"Yes, Diane, hold your body firm, not rigid. Let the music flow. Allow the music to guide your movement as I guide you through the dance."

They started off. A misstep here and there, but by the end of the dance. Diane felt better than she had ever felt before. Next came the Cha Cha. Now, this was fun. Lance had her making some advanced moves; sometimes, they missed a transition. They left the dance floor laughing.

Diane asked. "How long have you been dancing, Lance?"

“Since I was five years old. I once watched a competition with kids, and I thought it was cool. Then I saw Fred Astaire and Ginger Rogers in the movie *Swing Time*, which got me hooked. Of course, nowadays, everyone gets excited about *Dancing With The Stars*.”

Tango

A couple of minutes later. Lance said, “This next one is a Rhumba. Sit back; this is going to be a real treat.”

“Why?” Diane asked.

“Just watch Cody and Sabrina. You will see!”

Sabrina and Cody glided to the center, looking at each other. Their opening moves were full of power. The intensity of their moves, slow, halting, fast, once again slow, arms, hands moving, gliding, suddenly a push away and an equal drawback, faces just an inch away. Sweat was running down Cody’s face from the intensity of the dance until, finally, the last note finished.

“Oh, my God! That was amazing!”

“They do put on a great show. I think they call it dance ecstasy,” laughed Lance.

“Can we dance some more?” asked Diane as a swing was played. On to the dance floor they went. This time, Lance started with intermediate dance moves and found that Diane could do them easily. Soon, they were the center of attention as Lance made more complicated and more complex moves, even throwing in the 'pretzel,' the 'around-the-world,' through the legs, and even putting in a few Lindy-Hop moves. They both finished entirely out of breath, with huge smiles.

"You really know the swing, Diane. I hope you can come back to this part of Florida. You will be an excellent dancer if you keep dancing," said Lance.

The mixer was next, so Diane returned to where Cody and the others were.

"Hi Cody, Sabrina, Gwen! You guys dance great! Can you teach me to dance like that?" Diane asked.

"I'm not sure we will need to," said Cody. "You looked very good out there with Lance."

Diane grinned a little shyly. "You really think so? I thought I was awful."

Sabrina, Gwen, and Cody laughed at that comment.

"Really, for not having been to a ballroom dance, you were really able to hold your own with a great dancer. Lance won the US Open to the World Professional Rising Star International Ballroom Competition, where they compete in five dances: Waltz, Tango, V. Waltz, Foxtrot, and Quickstep," said Sabrina.

"What! Are you kidding? He seems so normal," said Diane.

They all laughed at that comment, including Diane.

The announcer was just completing a 50-50 raffle when he asked if anyone was new at the dance.

Cody immediately stood up and pointed to Diane. Diane blushed and tried to shrink out of sight. Cody had just finished telling everyone her home state when he took her hand and headed to the dance floor.

Some other new couples joined them on the dance floor, and soon, Waltz music was played. Cody took her right hand with his left. He put his right hand around her waist, held his frame, looked slightly left, and strode confidently into the dance, starting with his left foot. He had her whirling and twirling. They were on their second loop when all the other dancers came onto the floors soon; they were weaving in, out, and around through the many dancers on the floor. Skillfully, Cody made it look effortless. Just as they would arrive at a position where a couple had just been, there would suddenly be an opening, like magic. Other times, Cody would effortlessly reverse or rotate in such a way that there always seemed to be an opening. The dance ended with Cody putting Diane into an elegant layback with a low verbal command, so she positioned one arm gracefully above her head and her foot pointed.

"That was fantastic!" Diane exclaimed. "I need to catch my breath."

Cody led Diane back to the table, where Sabrina and Gwen clapped. Lance was also there saying, "Bravo, Bravo!"

The rest of the evening was filled with dance after dance until the lights came back on, and the magic was over.

Diane gave big hugs to Sabrina, Gwen, and Lance, who said they hope to stay in touch, perhaps on Instagram or Facebook, for future dancing.

On the return ride, the only conversation was when Cody asked Diane if she had a great time. She answered, "Wonderful."

ROCKETS TO THE MOON

9 a.m. Sunday, July 13th – It was 9 a.m. when Dr. Wood showed up for the drive to the Kennedy Space Center. They jumped in the car with Brooklyn in the front passenger seat and Thomas, Diane, and Dean in the back.

"Brooklyn, he was so wonderful! Our dances were so smooth. It was amazing," said Diane dreamily.

"Diane, was that your first dance?" Dr. Wood asked. "If so, be very careful. Much of what you experienced is show, especially with someone as skilled as Lance. He won the US Open at the World Professional Rising Star International Ballroom Competition. So, he is very, very good at dancing. Do not be fooled by his dancing that he has a romantic interest in you. It is a common mistake of newbies."

Brooklyn added in. "Listen to him. You may think Dr. Wood is just a professor, but he was often on the ballroom dance scene until he injured his leg."

"Really?" Diane asked.

"Yes, but that was several years ago. When I was pursuing my Ph.D. at Oregon State University, I helped form the university dance group 'Cool Shoes!'"

"Wow! What types of dances did you do?"

"They were all group dances. You learned a dance sequence and then you had to figure out how everyone could be in sync. We did two or three 1950s routines. Two 1940 routines and three 1930 routines. Let's see Jailhouse Rock, Rook-Around-the-Clock, and the Stroll for the 1950s routines. The 1940s routines were typical Artie Shaw and Glen Miller swing dances. The Lindy Hop was a dance from the 1930s, and boy, was that fast. Then, of course, we did a Latin and a Waltz medley, but enough on that, we are here."

They took a right and saw a couple of buildings and a bunch of rockets standing straight up.

"There's the rocket garden," said Dr. Wood. "I remember when this was a government operation. There were no fences and no money to preserve the systems. With the private sector running the Visitor Center, it costs much more, but they are preserving our space history."

They parked and headed to the entrance, where they purchased their tickets and took some photos.

"OK, let's head inside. MEET AN ASTRONAUT is in 10 minutes. We will meet one of Boeing's flight crew members who were on two space shuttle flights back in 2010," said Dr. Wood. "Personally, I prefer to hear someone from NASA's Orion capsule and the Artemis moon landing project and the possibility of soon flying to Mars."

Filing in, they found places down front so they had the best view of the screen when the astronaut put up pictures of Boeing's new spacecraft and talked about its various features. He ended with a video of fantastic scenes from his space flights and what his new missions were. They were very impressed and asked many questions.

Diane said, "Now, that is really cool, but I would go to the bottom of the ocean instead."

"Me too," said Brooklyn.

"OK, everyone. It is now time to hop on the bus for our tour. We will go by the assembly building where they put together the Saturn V rockets and all the space shuttle orbiters. As a kid, I got to go inside the building and watch Apollo 13 take off," said Dr. Wood.

The bus route took them by the buildings that once housed the assembly of the international space station modules. Dr. Wood mumbled how he saw every module as it was constructed and sent into space by the space shuttles.

He pointed out the now defunct observation building that overlooked the space shuttle launch pads and talked about the Boeing building where some of the current rockets are being assembled. As they came to the Space-X building and launch pad, he showed them the main thruster and two booster rockets that fly autonomously back to the launch pad.

"Now that's a thing of beauty, watching the booster land simultaneously on the pad," said Dr. Wood.

After touring Launch Pads A and B and stopping at the beach overlook, the bus headed to the Saturn V Rocket Building.

"Now, this is a real treat. What you are about to see is truly amazing! An entire Saturn V rocket was to be Apollo 18, but President Nixon canceled the flight and all future moon landings.

This rocket had been sideways outside in the open air for years, just rusting away. It was horrible. I stopped coming because I couldn't stand the thought of our heritage rusting away due to neglect. Luckily, when the commercial company took over the tourist aspect of the Kennedy Space Center, they recognized the importance of the rocket and restored it to how you will soon see it," said Dr. Wood. Just then, the big doors opened, and everyone filed in. After a brief video, they entered the actual launch command center of Apollo 8 from the 1960s.

"Of course, it wasn't originally here. It has been moved several times. When I was a kid, it was in those buildings that they recently used to assemble the International Space Station."

After watching the impressive launch of Apollo 8, the doors opened, and there was the Saturn V rocket!

"OMG! Now that is huge! Look at the size of each engine!" Thomas exclaimed.

"It is huge, especially lying on its side!" Dean said in amazement.

As Thomas and Dean zoomed here and there, Dr. Wood, Brooklyn, and Diane walked underneath the rocket.

Dr. Wood was reminiscing about an experience he had.

"Back in 2002, I saw an ad in a retirement newspaper for the elderly. There was going to be a celebration honoring the original seven astronauts at the 40th-anniversary celebration of astronaut John Glenn's flight around the earth."

"Wow! You were there!" Brooklyn exclaimed.

"Yes. It was a big dinner with all the astronauts attending, at least those still alive. The tables were set up under the Saturn V. The food was excellent. The butter was in the shape of space shuttles. The ice cream was shaped like a rocket. Walter Cronkite, who used to broadcast the launches, could not make it. That night, I met all the astronauts. I even got John Glenn and Gordon Cooper's autograph."

Dr. Wood sighed. "They are all gone now, but what an evening it was.

They went through the hall and all the exhibits. When they got to the lunar rover exhibit, Dr. Wood said as he looked at the vehicle, "Now, this was an experience for me."

"How so?" Diane asked.

"Well, I was standing about where I am when six old guys came up. They just looked like any old men you would see at an old person's home. But you should have heard them talk. You would think they were thirty years old. They were discussing every aspect of the moon buggy. They pointed out where all the main design problems were and how they solved them. This group of old men was the original design engineer of the buggy. I introduced myself and was immediately invited into their group. One of them asked me if I wanted to join them at their table, but another reminded him that it was assigned seating and that it was not possible. Great group of guys. I am sure they are all gone, too."

Brooklyn said. "What an excellent story. You need to document it in some way."

"Yes," replied Dr. Wood. "There are so many stories, like one of my conversations with a dance friend. I asked him how his week was. He replied that he had just spent the week changing a toilet. I said, 'What, it takes a week to change a toilet.' He promptly said: "So when was the last time you changed the toilet in the space shuttle Atlantis?"

Everyone laughed at that.

They finished touring the building, purchased gifts in the gift shop, and headed for the bus.

Dr. Wood said, "Do you see the buzzards over there?"

"Yes," answered Diane. "Boy, are they ugly."

"Well, NASA had a problem. You see, when a rocket or the space shuttle was launched, hundreds of vultures would take to the skies and could fly into or be hit by the rocket or shuttle. So, the engineers got to work devising a high-tech solution to eliminate the vultures. They tried everything, but nothing worked. One day, this little kid heard about their dilemma. He replied: 'Oh, that is easy to solve. Kill a cow and put it a mile or so away. Then all the vultures will go there, and you will no longer have the problem.' Well, they did it, and it worked! See, you guys can really help the world," said Dr. Wood.

They jumped on the bus and headed back to the visitor complex. They passed a huge eagle nest on the right side of the road. Unfortunately, the eagles were on vacation somewhere else.

They next visited the space shuttle Atlantis display at the visitor complex. Another case of 'Wow' that was heard by all was when the doors opened, and everyone saw the Atlantis for the first time.

"I saw this bird a lot," said Dr. Wood. "One of my best memories was watching a space shuttle launch from Titusville. I know this perfect site with very few people. Well, the launch was an evening launch on a full moon night. There I was, watching the full moon come up, when suddenly the evening sky lit up bright as the shuttle lifted off, and sure enough, the shuttle rotated right on time and cruised over the full moon! It was beautiful. I will remember that for the rest of my life! Fabulous!"

They left the space shuttle building and headed for the I-MAX theaters. After two spectacular short movies, they were worn out.

"Group photos in the rocket garden! Come on, this way," Brooklyn grinned and pointed to the right where the rocket tips were just visible.

Taking the photos was a little chaotic - everyone wanted to be in the back row. Thomas and Dean were pushing to get behind Dr. Wood and Brooklyn - Diane pulled them back to the front, and then Diane tried to hide. After watching the ruckus for a while (mind you, Dr. Wood was no help because he didn't mind being in front and striking a goofy pose), Brooklyn said with a phony and stern mummy voice: "OK, you guys, enough. The three short people stand in front." Finally, after getting a very nice attendant to take the photos, they returned to the car.

While they were walking, Diane quietly told Dr. Wood and Brooklyn. "This was really, really amazing. Thank you so much!" Then she cleared her throat, spoke a little louder, and suddenly appeared very grown up, like Brooklyn, "Now I think it is time to go back to our AUV."

They all laughed and talked about what they saw as they drove home.

CHECKLISTS

9 a.m. Monday, July 14th

"Good morning, everyone," Brooklyn said cheerfully as they met from breakfast.

"What's good about it?" Cody asked. "I got home at 3 a.m., and it is now 9."

"Well, it's not our fault you had to go night diving," Brooklyn replied. "Let's get to the project. Today, we will button her up, pressurize the pressure chambers, and put it in water to see if there are leaks. The chambers have already been tested without electronics to a 600-foot depth using the pressure chamber. Ok, let's get to it."

"I have a question," said Diane. "I noticed you have several of these bags?" Diane held up a couple of the little white bags that looked exactly like the ones one gets when buying shoes or other things. "What are they for?"

Brooklyn promptly asked, "So what do you think? Why are they in your shoe boxes and other things you buy?"

"I don't know." Replied Diane.

"Oh, come now. Think!" Brooklyn demanded.

"I guess those are to keep things dry," said Diane.

"Precisely! We place them in the pressure chamber where the electronics are. When you place one of these." Brooklyn put her hand on one of the vehicle's pressure chambers. "In the ocean or cold water, condensation will build up inside. Condensation will form water droplets that can drop onto an electrical circuit board and then 'pop' the board is dead. Dead board, dead vehicle! Dr. Wood was working on an ocean energy system a few years ago that was sponsored by a company. One of the company employees had an instrument open. The moment he plugged in the battery, a drop of sweat fell from his forehead onto the circuit board. Pop and a whiff of smoke arose - one dead board."

“OK, a dry bag for each pressure chamber,” said Diane. “Anything else?”

“Oh, yes, lots,” said Brooklyn. “Always run down and check off EVERY item on the checklist. We have that AUV there.” Brooklyn pointed to the Bluefin AUV. “Because someone in the Navy missed an item on the checklist, the pressure housing flooded. Poof, one million dollars out the window.”

“OMG!” Diane exclaimed.

Let's get at it,” said Brooklyn,

By noon, the vehicle was completely assembled; by 5 p.m., all systems were tested in the lab.

“OK, everyone. That is it for today. Tomorrow morning, we will first slightly pressurize the pressure chambers and put them in water to see if there is any leakage. If all goes well, we will start the vehicle and run through the checklist exactly as we would if we were in the field. If everything goes well, we will send it through various buildings, riding on a cart, pushing it as if running by itself through various obstacles. See you tomorrow,” said Brooklyn.

ψ

9 a.m. Tuesday, July 15th

"Hello everyone. Sit down, sit down. First, Dean and Thomas connect the pump to the titanium pressure relief valve and put a couple of pounds of pressure into it. Not much is needed. Diane, Derek, and Cody out back is a large plastic tank. Check to see if it is full of water and add more if not.”

Dean and Thomas unpacked the pump and set it up. Soon, there were eight pounds of pressure in the tank. Fifteen minutes later, Derek reported that the tank was ready. Cody, Derek, Dean, Thomas, and Diane picked up the vehicle and put it on the cart. “Damn, this thing is heavy! I can’t believe that it will be neutrally buoyant in the water. I would have expected it to sink like a rock!” Thomas exclaimed.

“Well,” said Brooklyn. “You just have to think about displacement.”

“Displacement? What is that?”

Brooklyn responded with a smile. "Well, this vehicle is about 36 inches long, 30 inches wide, and 18 inches high. It is shaped somewhat like a flattened symmetrical egg. Calculating the volume of the vehicle, we get around 44,000 cubic inches. Therefore, when we calculate the same amount of water it will displace when it is in the water, we can determine what its mass will need to be to equal the amount of water it will displace. I want you to tell me what you come up with tonight."

Cody positioned the AUV and cart close to the tank where there was a lift derrick. Connecting the AUV to the derrick, he lifted and rotated the vehicle over the tank. Everyone kept their fingers crossed for luck when it was dropped into the tank. One minute … five minutes … ten minutes. Nothing! Not a bubble.

"No bubbles!" Cody exclaimed.

"Is that good?" Dean asked.

"Oh, come off it, Dean! Of course, it is good. It shows that there will be no water intrusion," said Diane.

"Oh. You don't have to be all bossy about it," replied Dean.

"Knock it off. Let's get the vehicle back inside," said Cody.

They brought the AUV back into the lab where Brooklyn was waiting.

"OK, everyone. We just proved that the vehicle is watertight. So, what are we to do next?" Brooklyn asked.

Diane immediately responded. "We will review the checklist to ensure everything is operational.

"Excellent," Brooklyn responded. "Diane, you are the one with the checklist. You take control of the checkout. I am going to the store to get donuts, pastries, and coffee. Bye!" Brooklyn said as she walked out the door.

Cody responded quickly. "OK, Diane, you are in charge. Tell us what to do." He smiled, as did Derek.

"Ah, okay, ah, the first item says to visually check all components first before turning it on. Check connections and check for leakage of fluids, especially around the motors. Look for corrosion; look for anything out of the ordinary," said Diane.

"Quite right, Diane, but that is not the beginning. Fill in the time, place, test, or dive number. Write down all the details first. Such as who was at the checkout at the dive, etc.," said Cody.

"Really? At that detail?" Diane asked.

"Yes, those details are extremely important when something goes wrong. So, start over. I suggest using a blue pen instead of a black one. Therefore, if you receive a copy in black, it might have been changed from the original," said Cody.

Diane began again, writing down every detail requested on the checklist. An hour later, the vehicle was up and running. They even found three more items that should be added to the list.

A few minutes later, Brooklyn came with a large bag of goodies and coffee.

"Hi, everyone. Everything checked out, okay?" Brooklyn asked.

"Yes, Diane was excellent. At least for the first few checkouts, I suggest she perform each checkout with Dean and Thomas, verifying everything. She is really good at paying attention to detail," said Derek.

"Excellent! Everyone. Select a pastry that you like and grab a cup of coffee or juice. Today will be a long day of testing. We will take breaks two at a time for lunch. I want to keep the vehicle running continuously for hours to break it in and verify all systems and sensors," said Brooklyn.

Thirty minutes later, they were once again gathered around the vehicle.

"OK, lift the vehicle back onto the cart. Make sure to place the marine batteries on the cart's bottom shelf," said Derek.

The four guys picked up the AUV and placed it on the cart so the lasers would be free to rotate.

"Let's go to the aeronautical building. It has a weird shape," said Brooklyn. "I have the vehicle in partial simulation mode. It thinks it is turning the motors on and off, and I give it false depths."

"How fast should we go?" Cody asked.

"Normal walking speed should be fine," Brooklyn replied.

All six of them were off. Soon, two of Brooklyn's friends showed up, wondering what was happening. Then, one of Cody's friends, three of Derek's friends, and friends of friends were all parading across campus. Soon, security showed up. Then, as the horde entered the building, the Dean of Aeronautics joined in with the throng as they went down the hall into and around each classroom. People were everywhere. It was almost a festival. As they left the building, the university's president came up with a smile.

"I heard there was a commotion over here. Something about autonomous vehicles mapping the university: I first thought it was Google. I did not expect it to be our ocean engineering department! Perhaps civil or mechanical engineering. Shouldn't you be doing this underwater?"

Diane, who was closest to the President, said. "We are making this AUV's first flight. We just laser and sonar mapped the aerospace building, and now we are off to the engineering building to map it."

The president looked impressed. "Are you one of Dr. Wood's Ocean Engineering freshmen students?"

"Yes, we are working with Dr. Wood, but no, I'm in middle school, as are my two friends here, Dean and Thomas," said Diane.

"What! Middle school? That is excellent!" The president replied.

Just then, Brooklyn, Cody, and Derek came over.

"Hi, Mr. President; yes, let me introduce our team. Here are Diane, Thomas, and Dean. I am Brooklyn; this is Cody and Derek. I am a master's student working on my thesis under Dr. Wood. Cody and Derek are juniors about to start their senior design project, and these three are visiting this summer to learn about AUVs," said Brooklyn.

"Wonderful. How about mapping my building and office?"

"Sure," said Brooklyn. "Everyone. Change of plans. We will map the president's office before we go to the engineering building."

As they headed in the opposite direction. The President got on his cell phone. Diane overheard him say: "Melanie, get Adam with a camera crew to my office, NOW! Immediately! We will enter the building in five minutes with a horde of students and another of Dr. Wood's devices."

Just as they arrived at the building, Adam and a cameraman came running up, sweating in the hot sun.

"Quick, Adam, get that camera running and take many photos!"

Diane led the way into the building, ensuring no one was in the way of the lasers as the vehicle was taken room-to-room. When they finished the tour, the president met them at the door: the cameraman and Adam at his side.

"Hi Brooklyn, Diane, and the rest of Dr. Wood's team. Come into our conference room. Brooklyn, could you sit there and Diane next to her, please. Excellent, and if the others could fill in the spaces across from them. Perfect. Adam, I want your cameraman to film these wonderful researchers and me in this interview. OK, is everyone set? Great. Now, Brooklyn, start from the beginning and tell us about the project, who your colleagues are and how they got involved, and what you plan to do with your fine vehicle in the lobby."

For the next hour, Brooklyn, Diane, and Cody explained everything. The president asked question after question. Close-ups and group shots were made when everyone got in on the discussion. The interview ended with all of them promising to keep the president and Adam with his cameraman in the loop as they continued the project, especially when they deployed the vehicle.

ψ

9 a.m. Wednesday, July 16th

"Wow, what an exciting day it was yesterday - meeting the president of the university and the Dean of Aeronautics. Did you know that he was also an astronaut? He was on three space shuttle flights," said Diane.

"Any idea where Brooklyn is?" Derek asked. "She is never late, and it's now past 9 a.m."

"No," replied Cody, "I assume she spent the night running through the data."

"So, what should we do today?" Diane asked no one in particular.

"Did anyone recharge the batteries?" Cody asked almost simultaneously with Diane.

"Yes, the batteries are ready to go," said Dean.

"Perfect, we must perform power output tests and longevity studies at various velocities. After we do our checkout procedure, we will use the mounting bracket that Brooklyn 3-D printed to mount the vehicle in the tank. Then, we will run the vehicle at 50% velocity until its batteries are down to 20%, and we will keep the data acquisition cable attached to monitor all the vehicle parameters. The lasers and sonar will always be active, as will all the other sensors. If we have calculated correctly, the vehicle should run continuously for five hours," Cody said.

Derek continued. "We saw yesterday it took Diane two hours to complete the checklist. So, we should be able to start at 11:30 this morning."

"Now that Diane, Thomas, and Dean are professionals at setting up the vehicle, they should be able to do it in an hour," said Cody.

"But, but," said Diane.

"You know what to do. I am sure Thomas or Dean can help you if you forget something. Here is my cell phone number if you get in a jam."

Directing his gaze at Derek, Cody continued. "Derek, I want to check out the new scooter that the dive shop just got in."

Both Cody and Derek left the lab and headed to their car.

"Ah, can you believe it? They just walked out the door and left us with all the cleaning!" Dean exclaimed.

"Well, we know what to do. Thomas, please tell me what you see concerning the vehicle. By the way, we need to name her. I suggest we call her *Nautilus* after Captain Nemo's submarine in 20,000 Leagues Under the Sea," said Diane.

"Yes. That is a perfect name," said Thomas.

"I thought we should call it 'Fetch' since we would send it after things. But I guess everyone would prefer Nautilus."

"No! I don't think it should be either of those names. Just look at it! What does it look like?" Diane asked.

The others just stared at the vehicle.

"What is wrong with you guys? It looks like an OREO cookie!" Diane exclaimed with disbelief.

"You know, she is right. It looks like an OREO cookie. Okay, we will call it the OREO cookie," Thomas and Dean nodded in agreement.

Diane, Thomas, and Dean went through the checklist. With all systems checked out, they sat back, and in walked Cody and Derek.

“What? It is only one hour, and you are sitting back and relaxing! It would be best if you were well underway with the checklist. Haven't you even started?” Cody asked.

“Ha, Ha, Ha, here is your checklist, and if you had looked at the computer screen, we ran all the tests Brooklyn said we should. Everything turned out perfect.” Diane replied indignantly.

“Wow!” said Cody, who gave Derek a wink. “Oh, you missed an item!”

“Where?” Diane demanded.

"Right here, you forgot to sign your name.”

“Good job guys. Cody is just upset that he lost the bet. I said you would finish in under an hour, and he said that it would take you a minimum of an hour and a half,” laughed Derek as he put out his hand as Cody handed over his SUV keys. “Excellent, now I have a nice car to take Johanna out for the murder mystery I have arranged for tonight.”

"OK, let us get back to work. Shut the system down, and let's mount it to the tank,” said Cody.

Derek pushed over the transport cart, and Thomas unplugged the vehicle from the power outlet.

They packed what was needed in a box. The test tank mount was attached to the vehicle and rolled out to the test tank. It took the five of them to mount and lock it in the tank. Of course, when they started it up, they immediately received a system error.

“Hey guys, there is something stuck in the left horizontal motor. It looks like a set of keys,” said Diane.

“Derek, if you cannot keep track of my keys, you cannot use my car bet or no bet; you’re just going to have to take her on your motorcycle,” said Cody.

“Damn! She will not like it if it rains,” said Derek.

Soon, all the lights turned green. The vehicle was ready for her tests.

“I know this test will run for hours, but someone needs to be here if something fails and it doesn't shut down. Okay, Diane, you get the first hour, Thomas the second, Derek the third, Dean the fourth, and

I will take the fifth. If more time is needed, we will all be here after my term," said Cody.

The test started, and the first hour ran smoothly. During the second hour, the internal temperature climbed, but once it approached 'hot,' the internal fan turned on, and the temperature leveled out still in the green zone.

Hours later, everyone met in the lab.

"Six hours and twenty-three minutes and the batteries finally reached 20%. Excellent!" Diane exclaimed.

"Wow! You sounded like Brooklyn when you said that," Cody said, impressed.

Right behind him, Brooklyn said. "Of course, she did. Brilliant minds think and sound alike. Unlike others, I know."

"Brooklyn, where have you been? Is everything all right?" Diane asked.

"I was up till dawn scanning the data from yesterday. I went to bed at 7 a.m. I got up a while ago and had breakfast, lunch, and dinner," said Brooklyn.

"We just finished a six-and-a-half-hour test running all systems and sensors with the vehicle at 50% speed. All the data was downloaded 15 minutes ago. I made a copy. Here it is on this USB thumb drive," said Derek.

"Excellent. Thanks. Shut everything down, run the post-operation checklist like last night, and put everything away. Please do it again tomorrow, but cycle from 20% speed to maximum speed every 10 minutes until the batteries reach 20%. By the way, don't forget to recharge the batteries tonight," said Brooklyn.

"Sure, Mom. Do I have to brush my teeth, too?" Cody mocked.

Diane, Thomas, and Dean smiled.

"See you tomorrow evening and make another USB thumb drive with all the data. Bye," said Brooklyn and walked out the door.

"Wasn't that strange?" Diane asked.

"No. Not at all. She is like that when things are going well, and everything is going well. OK everyone, let's finish and get some food and sleep; tomorrow will be another long day."

ψ

9 a.m. Thursday, July 17th

"Hi everyone," Cody said.

Diane was already running down the checklist with Dean.

"Diane, where is Thomas?"

"The last time I saw Thomas, he was going back to order two pancakes with blueberries, and that was after he had already eaten an omelet and a bowl of cereal," said Diane.

"OK, I think he'll be along shortly," replied Cody. "Let's get these tests started."

As they did the day before, they worked steadily throughout the day, monitoring the vehicle until it shut down when the battery power reached 20%.

"Excellent! What a day! And here comes Brooklyn."

Brooklyn entered the lab looking quite sleepy.

"Hi Brooklyn, we just had a very successful day. Here is the USB thumb drive with today's data," said Cody.

"Hi! How is everyone?" Brooklyn asked.

"Fantastic!" Diane replied enthusiastically. "Thomas, Dean, and I ran all the tests by ourselves! Of course, Cody and Derek were there if anything weird happened, except when they were at the dive shop checking out some new scooter!"

"Ah hem, you did what?" Brooklyn asked as she turned and looked sternly at Cody. "You know you weren't supposed to leave the lab!"

"Yes, but Diane had everything in hand and... How else do they learn to stand on their own two feet? Anyway, we did not leave to check out the scooter. We went to Ed's office two doors down and kept track of what was happening in the lab." Cody replied.

"So, how did you do that?" Brooklyn demanded.

Cody looked over at Derek, who promptly said. "Well, I rigged four video cameras and a mic there and there." As he pointed to various tiny cameras that were barely visible." Then I put a monitor in Ed's office so that we could keep track of things."

"You have been spying on us!" Diane spurted out indignantly.

"Only because we were told not to leave, and it was the only way we could think of to allow you the freedom to take the lead. Otherwise, you would have always turned to us for solutions. Sorry," said Cody.

ψ

9 a.m. Friday, July 18th – Diane, Dean, and Thomas came into the cafeteria with wet hair and plopped their stuff down at the table Brooklyn, Cody, and Derek were sitting at.

"Now, that was wonderful. A quick swim before breakfast. I wish our school at home had an outdoor swimming pool," said Dean.

"Oh, no, you don't. It is 5 degrees Fahrenheit in the winter and cold outside for the entire school year at home. Perhaps if they had a heated inside pool where the current gymnasium is, I would say that is perfect," replied Diane.

"Very nice, now go get some food. We have a lot to do today, and I want to discuss what everyone will be doing," said Brooklyn.

"Oh, Brooklyn is in a mean mood today," laughed Cody. "You better do what she says, or she will bite your head off."

Brooklyn grinned. "Just go get some food. After yesterday's stunt, I'm only going to drown Cody and Derek this afternoon if you want to watch. Other than that, it should be a pretty fun day."

Diane, Dean, and Thomas headed to the grill for pancakes, French toast, eggs, bacon, grits, and fresh orange juice.

"Now, this is one thing I love about Florida. You can get fresh squeezed orange juice, grapefruit, and passion fruit juice. Then there are the fresh mango slices, fresh pineapple, and avocados. Yesterday, I saw some ripe bananas in the botanical gardens. I think I will snag a few on the way back to our dorm," said Thomas with mischief in his eyes.

"No, don't get into trouble," said Diane. "We have a lot to do and cannot afford the time to bail you out of jail."

When they returned to the table, they found Brooklyn, Cody, and Derek deep in conversation about the previous day's results.

"Hi everyone; as you dig into your food, let me say you three have done a wonderful job over the last few days. Tomorrow and Sunday, you have off to do whatever you want to do. Today, even though Cody and Derek don't think it is necessary, we will run through the complete set of tests that we have just completed. Things are going too well. I don't like it when things go too well. That means something will go wrong later when we don't want them to,' said Brooklyn.

"Oh Brooklyn, you worry way too much," said Derek.

"Brooklyn, what changes are you thinking of," asked Diane.

"Not many, just increasing and oscillating the speed of the vehicle. Provide it with obstacles that will cause it serious difficulties, and so on. If the vehicle can handle those, we will put it in the pool on Monday," replied Brooklyn.

So, that was their day. Putting the vehicle through every test they could think of. Occasionally, a glitch appeared, but nothing 20 minutes of code updates couldn't fix.

The day dragged on, and by 5 p.m., they were dirt tired. As they were about to leave, Brooklyn announced. "OK, see you tomorrow at 9 a.m. Right here in front of the lab. We are going to Sea World."

"Really! Excellent!" was everyone's response.

What an exciting week this had been, and now off to Sea World. Diane was in heaven!

SEAWORLD

9 a.m. Saturday, July 19th – Sea World!

“Hi Brooklyn,” said Diane.

“Hi everyone. Who is driving?” Brooklyn asked.

“Derek is. Cody doesn’t want to miss dancing tonight, so we will not be going,” said Diane. “I would have loved dancing again, but I have never been to Sea World.”

“Derek parked his car across campus and should be here soon. Or so he said,” said Dean.

Five minutes later, as sweat beads accumulated on everyone’s faces, Derek drove up in a Mercedes.

“Hi Derek. Cool car. You look so domestic in your car,” Brooklyn said.

“It’s my parents’ car. It rides really sweet and quiet, and my dad put in an excellent stereo.”

“Shot Gun!” Thomas cried.

“I don’t think so. You are in the back with us,” said Brooklyn.

One and a half hours later, they pulled up to the entrance after speeding down the 'Beach Line.’

“OMG! It sure is expensive to park!” Diane exclaimed.

“Yes, but it is still cheaper than the other theme parks in the area,” said Derek. “Anyway, use my season pass. Parking is free with it.”

Soon, they were off to the parking lot.

“Jeez, we are way out in row 23. Are we going to have to walk far?” Thomas asked.

“No, the entrance is right there. We are quite close,” said Brooklyn.

As they approached the entrance, Thomas said to Dean." Your entrance is right there.” Pointing to the small building just to the right of the entrance. The building allowed the public to leave their dogs and cats while visiting the park.

"Ha, Ha, Ha," Dean snickered, "I may just stick your head in a cage with a rabid cat!"

Everyone passed through security and waited in line for tickets. Then, they had identification pictures taken and their entrance cards scanned.

"Boy, this is getting ridiculous! All of this to get into a park. You would think that they could come up with something simpler," said Thomas.

They picked up a map and went around the corner to meet the newborn dolphin in the dolphin nursery.

"OMG! Is she cute or what? She is so small. Look how smooth and sleek she is. It is wonderful to watch baby and mama swim by," said Diane.

Dean immediately chimed in. "Look, Diane, over there is another mom and baby, but she is definitely older."

"Quite right," said one of the SeaWorld employees. "She just turned nine months old last week."

"Neat," said Diane.

"Do you know that dolphins dive up to 1,000 feet and can live up to 50 years," said the SeaWorld employee.

"Wow!" No, I didn't," said Diane, while Dean shook his head.

"Dean, we are continuing to the dolphin feeding area," said Derek.

"OK," said Diane, turning to the employee. "Got to go. Perhaps we can talk more later."

"Hey everyone, it is almost time for the dolphin experience! We can feed, touch, and play with a dolphin from the side of the pool. Let me get tickets," said Brooklyn.

"No, I'll pay for Thomas, Dean, and myself," said Diane.

"Nope," said Brooklyn. "This isn't my money. This is from Dr. Wood's money that he got from a grant."

"What is a grant?" Diane asked.

"A grant is money given to a researcher by the federal government. You see, the government, in this case, the NSF or National Science Foundation, is interested in getting the younger people, especially women, interested in Science, Technology, Engineering, and Math, or in other words, STEM."

"So that's what 'STEM' means," said Diane.

"Yes, you see, Dr. Wood wrote a grant to design and develop an autonomous underwater vehicle to solve science issues while educating all of us, especially you, Thomas, and Derek. So, all your expenses, except souvenirs and miscellaneous items, are paid for by the grant: sea turtles, Kennedy Space Center, Sea World, vehicles, your housing, food, and transportation."

"Wow! Thanks. I had no idea."

"Yes, so don't worry about money unless you're buying a giant seal baby or penguin," said Brooklyn.

The next hour was spent getting their fingers stinky with dead fish as they fed their dolphin. They all got to touch and interact with a dolphin and, of course, lots of pictures.

"Brooklyn. I didn't want to say it before the others, but I got to swim with the dolphins in the Florida Keys last spring. My best friend Bronwyn and I were researching global warming, ocean and coastal water warming, and sea-level rise and its effects on coral in the Caribbean. My whole family and Bronwyn started at Epcot Center and ended up snorkeling the Wind Jammer wreck in the Dry Tortugas," said Diane.

Brooklyn looked at Diane with astonishment. "Wow! To the Dry Tortugas. I also dove the Wind Jammer wreck with Dr. Wood just last summer. We video documented the entire area and made a 360-degree video showing the entire wreck using virtual reality glasses. When we get back, I will show you."

Thomas came bounding over. "Brooklyn, Diane, we have just enough time to ride the 'Manta' rollercoaster before the dolphin show starts!"

"All right, let's do it. I want to show you the dragon seahorses," said Brooklyn.

"What are dragon sea horses?" Diane asked.

"You know what sea horses are? Right?"

"Yes, of course, they are fish that look like horses that live in areas with a lot of seagrasses." Replied Diane.

"Exactly, and dragon seahorses are very exotic seahorses that look like seaweed."

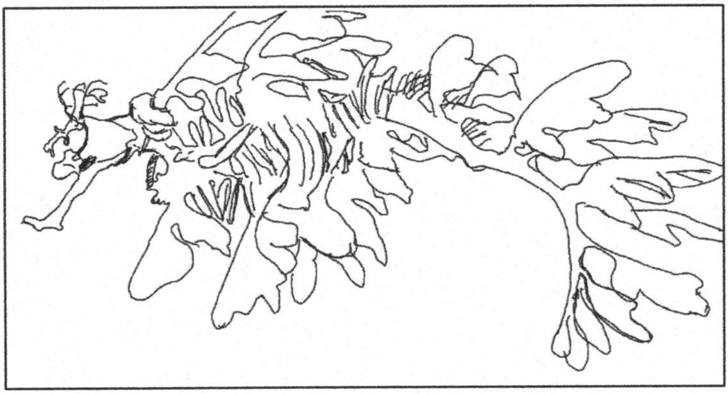

Diane's Sketch of a Leafy Sea Dragon - Seahorse

Before entering the 'Manta,' they emptied their pockets and put their sunglasses and all loose items in a storage locker.

"You are going to be glad you put everything in the locker. This is a wild ride!" Derek exclaimed.

They passed several aquariums, one with spotted rays and another with large exotic seahorses called sea dragons.

"Wow! They are amazing. I didn't know anything like it existed," said Diane.

"Yes, and look at the others. Some are more purplish in color," said Brooklyn.

"Let's go. The line has moved," said Derek.

"Damn, that looks intense. Look how they strap you in," said Dean as he proceeded to his spot. They were in one line. Locked in, they were rotated into position.

"OMG! I'm not sure I want to do this," said Thomas.

"Too late now," said Dean as the coaster moved forward. Leaning heavily on their legs, hips, and chests, the coaster moved higher and higher, and then over they went, whoosh as a manta might do flying through the ocean.

Screams were heard everywhere. Thomas had his eyes screwed tight. His teeth gritted together. Diane was screaming, 'Weeeeeeeeee!' Over, around, loop-the-loop over the water as fountains of spray filled the air. Then whoosh, clunk, clunk, clunk, as everything stopped.

Diane was shouting. "We must do that again!"

"OMG. Never ever again am I getting on this or any other roller coaster," said Thomas.

"Ah, don't be such a wuss," said Dean. "It wasn't that bad."

"Oh yes, it was. I think I am going to be sick."

"No, you're not, and anyway, there is no time for it. The dolphin show starts in ten minutes. Everyone, let's go," Brooklyn said excitedly.

They took a right and then left, and there was the entrance to the dolphin show.

"Everyone, to the middle. Quickly now. Make sure you are a little left or right of center," said Derek.

"Hey, it says that this is the splash zone. Are you sure this is where we should be?" Thomas asked.

"Yup, and if we are lucky, we will get wet," said Brooklyn.

"Quickly, it is about to start."

And yes, they got wet, as one of the dolphins playfully sent a whole tail full of water at them.

"I loved that they focused on the environment and how the activities used to train the dolphins are what occurs in their natural environment," said Diane.

"Yes, me too. When I first came to SeaWorld, I was worried about animal abuse and unnatural activities for the dolphins and whales, but all my fears were put to rest. Especially their rescue programs. Did you know that just this year alone, they rescued 43 manatees?" said Brooklyn.

"OK, what is a manatee?" Diane asked while Thomas and Dean nodded their heads in wonder.

"A manatee is sometimes called a sea cow. It is related to the elephant but lives in the coastal and estuary waters of Florida, the

Caribbean Islands, and the Gulf areas down to Mexico. OK, OK, you guys need some education. This way everyone, we will check out the manatee tanks," said Brooklyn.

Passing by the baby and juvenile alligators, they rounded the bend and came across a deep pond where at least ten baby, juvenile, and full-gown manatees were.

"Hey, everyone. Over here, we can view them from underwater," said Dean.

They made their way down below, where they had a spectacular view.

"Look at that tail. Huge! I was canoeing with my boyfriend last year on Turkey Creek. We were just resting and having a bite to eat when suddenly the boat moved. It was a manatee coming for air and to say hi. It rolled on its back for us to rub its belly," said Brooklyn,

"You got to be kidding," said Thomas.

"No, they are very gentle creatures. Well, we rubbed its belly for about a minute or so. It rolled back, and we thought we should continue when something suddenly spooked it, and a wave of water covered my boyfriend. What a shot! I couldn't stop laughing. I think the manatee knew something since we broke up a few weeks later," said Brooklyn with a smile.

Continuing, they made their way to the Kraken roller coaster. Brooklyn, Diane, and Thomas opted out and instead went on the Journey to Atlantis water ride. Everyone came out alive and headed for lunch. It was already 2 p.m., and they were very hungry.

"I know a great place for ribs," said Derek, so they had ribs, chicken, sausage, fries, and cokes for lunch.

"Hey everyone, so what next? Penguins? Sea Lions? Sharks? Killer Whales? Beluga Whales?" Derek asked.

After a quick vote, they headed to the shark exhibit.

"Hey everyone, it says: 'there are more than 350 types of sharks throughout the world,' and it says: 'more than 100 million (100,000,000) sharks are killed each year!' Wow! We kill that many per year! That is unbelievable," said Diane.

"Yes, most of the killing is done for shark fin soup, which is a delicacy, so they capture the shark, chop off its fins, and then release them to die an agonizing death," said Brooklyn. "Here it is on this sign. Seventy million are killed for soup."

"That is awful! What can we do about it?" Diane asked.

"Tell your parents about it. Get them to write to your members of Congress. Get together with your classmates and write letters. The key is to get the word out," Brooklyn said.

"Hey guys, read this: 'Long before dinosaurs, long before birds, long before humans, there were sharks,' now that's cool," said Thomas.

As they exited the exhibit, sharks were on their minds. They spent another half hour viewing the leopard sharks, brown-banded bamboo sharks, white-spotted bamboo sharks, and white-tipped reef sharks,

> Did you know sharks have keen olfactory senses, and some species are able to detect blood with as little as one part per million in seawater?
>
> Did you know the whale shark is the largest fish in the world with a length of up to 39 feet?
>
> Did you know that most modern sharks can be traced back to 100 million years ago?

"Everyone, it is time to go to the whale show. I would say we sit at the same relative position as with the dolphins," said Brooklyn.

"Could we sit a little further back? I think a whale can make a much bigger wave than a dolphin," said Diane.

Everyone agreed and headed to the whale show.

"Beautiful!" Diane said as the whale slid back into the water." They are so beautiful. It is hard to believe that some people don't want us to see these magnificent animals."

"Don't blame them too much. They are trying to do what they think is right – to stop exploiting animals. They forced SeaWorld to clean up their act."

"OK, I didn't know that. Hopefully, they now know that these shows make people aware of their beauty and may act by voting for environmentally friendly public officials," said Diane.

"OK, everyone. It is time to head home slowly. Any last wishes?" Derek asked.

"Could we check out the 'Wild Arctic' shipwreck? It says they have a couple of Beluga whales, walruses, and seals?" Derek asked.

"Excellent suggestion," said Brooklyn. "Do you want to hike or take the helicopter?"

"Of course, the helicopter," said Thomas.

"Okay, let's get in line," said Derek.

Flying in was eventful, with everyone bouncing around in the simulator as if they had a storm behind them. They saw caribou, polar bears, and seals on the flight in. Finally, arriving at the base, they walked the final yards to the entrance, where they came to an upside-down ship from the 1840s. Two beluga whales were swimming in the waters next to the ship. On the other side of the ship were seals and walruses, which you could view above or below water. It was so realistic that you felt like you were really up north in the Arctic.

"Fabulous, absolutely fabulous. I want to join an expedition to the Arctic or Antarctica," said Diane.

"Don't be daft. Kids our age do not get to go on adventures," said Thomas.

"Actually," said Brooklyn. "I know this kid; his name was Sontash, if I remember correctly. He had just finished his first year of college, and he was invited on a cruise that left New Zealand and went to a couple of spots in Antarctica. He ran the SideScan SONAR."

"Wow! Really!" Said Dean.

"What is a SideScan SONAR?" Diane asked.

"A SideScan SONAR typically looks like a torpedo that is pulled by the ship. The torpedo sends sound waves that bounce off the bottom and return, allowing the operator to see what is on the bottom of the Ocean. Sontash found two ancient shipwrecks in excellent condition," said Brooklyn.

They all headed to the park exit. Diane, Thomas, and Dean were in total amazement at all the possibilities of what they could do and be when they grew up.

ψ

Noon Sunday, July 20th – Panther Dining Hall

"Hi everyone," said Brooklyn as she joined Diane, Thomas, and Dean. "I trust that you found the All-Faith Center for whichever service you were inclined to attend."

"Yes, thanks for telling us about it. We missed the last couple of weeks because so much was happening. And today, it is great that we can do whatever we want. I am all for hanging out at the pool most of the day, relaxing, and perhaps going to the beach in the evening. Maybe we will even see some hatchling Loggerhead Sea Turtles emerge and head to the sea. Even if we don't, the full moon will be spectacular," said Diane.

"Oh, I forgot to remind you all. Don't forget to stay up until 1:30 a.m. tonight. You might remember that it was announced at Kennedy Space Center that SpaceX would launch its manned rocket. It should be brilliant. In fact, if you are already going to the beach, I would say stay until after the launch. Night launches are fantastic," Brooklyn exclaimed with a big grin.

"Really? We forgot, but how can we return from the beach so late? We were going to go with another student who was going with his girlfriend to watch the ocean as the sun sets," asked Thomas.

Brooklyn thought for a minute and finally said. "OK, okay, you three come with me tonight. I will swing by around 6 p.m. Then we will go for a pizza on the beach for dinner."

"Oh, thank you, Brooklyn! You are the best!" simultaneously exclaimed Diane, Thomas, and Dean.

The rest of the meal, they talked about rocket launches, their trip the previous week to the Kennedy Space Center, and Dr. Wood's stories about the old Space Shuttle launches that he watched from Titusville, especially how he watched the space shuttle Endeavor launch, which arched over a full moon that had just come up.

The rest of the day was spent frolicking in the university pool and sunbathing, although it was too hot.

At 6 p.m., Brooklyn drove up, and they piled in with towels to lay on while waiting for the SpaceX launch.

"Mm-mmm, now that was a good pizza. Maybe not as good as back home, but it was pretty good," said Dean.

“My Calzone was excellent,” smiled Thomas as he wiped the sauce off his face and finished his Coke.

“Hey, everyone. Let’s walk the beach and see if we can find any loggerhead baby turtles coming out of their nests,” said Brooklyn.

Everyone excitedly headed to the ocean. Splashing their way down the beach, the evening turned into night, and they ended up watching the stars and counting the satellites going by. It was one of those perfectly clear nights where you could see for miles. Around 1 a.m., the beach started to fill up with people to watch the launch. Someone nearby said loudly. “The launch is 15 minutes and counting. Everything is going smoothly.”

“Wow! This is really cool! I have wanted to see an actual launch since I was a little kid. I wish I could have seen a space shuttle launch. My mom says they were amazing,” said Diane.

“10 – 9 – 8 – 7 – 6 – 5 – 4 – 3 – 2 – 1, and we have lift off,” came from a nearby cell phone.

“Where is it? Where is it?” Cried Thomas.

“There it is! No, over there! You are looking too far east. The coast bends around,” said Brooklyn.

“Wow, it looks like a giant Roman candle shooting into the sky,” said Thomas.

“In a minute, look for the separation of the boosters. After they have separated, you will see them fire their return-breaking rockets. They will land back at the launch site. We will not see the main propulsion thruster as it falls. It is usually far out at sea,” said Brooklyn.

“There is the booster separation,” someone yelled on the crowded beach.

Dean yelled, “There are the two boosters; I see their rockets firing!”

“Now that’s amazing!” Diane exclaimed.

“The SpaceX rocket is now almost out of sight. All I can see is a tiny point heading northeast,” Dean reported.

A couple of minutes later, after watching the returning booster rockets make a series of slowing burns, the two booster rockets came down exactly together. They slipped behind some buildings at Cocoa Beach, 30 miles north of where they watched the launch.

The SpaceX rocket was now so small that your eye could hardly keep it in focus. Finally, it was gone, and people started leaving the beach and heading home.

“All right guys, time to go back. Don’t forget to tell your parents tomorrow,” said Brooklyn as she picked up her beach towel, and they all returned to the car.

DRUGS, SUBS, AND BUGS

In the Bahamas, two hundred miles to the east-south-east of Melbourne, Florida, Luke, a submarine pilot was saying:

"Miguel, sir, the sub is loaded and ready for departure. Each scuba tank is 80 % full of cocaine packaged in plastic bags and pressurized to random pressures between 250 to 800 pounds per square inch, just as if the tanks were used in a real dive."

Miguel, second in command of the 'Bahama Mamas,' a local drug cartel with links to a Columbian cartel, has been smuggling drugs into the United States for several years. He had overheard a professor talking to his graduate students at a bar in Key West about when the harbor police pulled his boat over and checked everything except the scuba tanks. They would only get air from the pressurized tanks even if they had. It was such a clever idea, but how to make it sustainable? That is when Miguel had his great idea of using a tourist submarine to transport the drugs underwater to a location off Miami Beach. For over two years, the scheme had worked flawlessly, and now, what to do with the two cave-diving intruders? We cannot hold them forever. Kill them? That will be up to Mr. Rodrigues. He is the one who must make the tough decisions. But now, let's get the shipment underway.

"In you go, Carlos and Luke. Take the next shipment. When you return, Mr. Rodrigues may have another assignment for you," said Miguel.

"Sure thing, boss," replied Carlos, climbing onto and into the sub. Luke followed closely behind and pulled the hatch closed. Twenty minutes later, the submarine sank out of sight.

Miguel headed to Mr. Rodrigues' office near the surface of the complex. The complex was an underground submarine base on Abaco Island, which had been built by a Chinese company several years back. Mr. Rodrigues, with the advice of this second in command, purchased the area and expanded the underground complex to include an office, living space for ten people, storage space for drugs and scuba tanks, and docking space for one tourist submarine which transports the drugs from the island to Miami, Florida.

As Miguel entered the office, he heard Mr. Rodrigues' number one henchman, Carlos ask, "Mr. Rodrigues, what are we going to do if the drug police find us out?"

"Well, first we escape in one escape submarine, which, with its fuel cells, can reach anywhere in the world, but I know that is not the real question. You want to know how we will continue our smuggling operations. Well, a professor at the Florida Institute of Technology, Dr. Wood. He has some remarkable graduate students. They came up with the design of a vehicle, which is a plane that can transition into a boat, then into a submarine, and back again. This means we can load it with cocaine and send it as a submarine to the coast, where it will become a boat that will travel up a river well inside the USA. Next, it will transform into a plane and fly to one of our many pick-up areas throughout the state." Rodrigues finished with a big grin.

"But Mr. Rodrigues, how are you going to get the plane?"

"Now, that was the easy part. All I had to do was to pose as a standard entrepreneur with some money to spend on radical ideas. Professors are always in need of money to keep their labs going and to support their graduate students. So now we have a patent and are building our first prototype," Mr. Rodrigues replied. "Can you believe it? I am on a patent for ten percent. Now, what a pathetic return on investment! I cannot believe professors sell themselves so cheaply. I may have to really give this guy some real money so he can really invent something useful."

"Mr. Rodrigues, just to let you know, the sub is on its way with the next shipment," said Miguel.

"Thanks, Miguel. Can you send the prisoners to me? I need to find out who they are and what they were up to," said Mr. Rodrigues as he puffed on a Cuban cigar.

Ten minutes later, a guard brought Rob and Leanne into the office.

"Please have a seat," said Mr. Rodrigues when the door was closed. "I have a few questions to ask you." As Rob and Leanne sat, Rodrigues started, "So, who are you two, and why are you here?"

Rob and Leanne looked at each other and realized it was pointless to make anything up and promptly filled Mr. Rodrigues in on why they were there. Especially about what they knew about the story they had heard in town and how they wanted to dive the blue hole before it was destroyed.

Rob continued, "This facility is amazing! How were you able to build it without anyone finding out?"

"Ah, now that is the simple part. We didn't; the Chinese company did. They tried to slip one by the Americans and set up a secret agreement with the Bahamas government to build a container depot for shipping goods to and from China. Except they would build a submarine base right here, just a mere 100-plus miles from America," replied Mr. Rodrigues. "Can you believe it! They almost succeeded, except for this dive boat during the Chinese New Year, totally by accident. Usually, there were many guards to fend off unwanted visitors, but on that day, everyone was celebrating. The dive boat was looking for one of the blue holes that were all over the island. And boy, did they come on an island in transition. YouTube is what killed the Chinese company's plans since the people on the dive boat were really upset that one of the best dive areas of the Bahamas had been destroyed. They took videos and pictures and plastered the Internet with all the scenes. Someone at the CIA must have caught wind of it. The place was swarming with government officials one day later, and the Chinese company was gone a day after that!"

"And the submarine tunnel?" Leanne asked.

"Yes. The company did a fine job. Look at the workmanship! We barely had to do anything at all," continued Mr. Rodrigues.

"Mr. Rodrigues," continued Rob. "Do you realize that if you use some of your money to buy this section of the island, you could build a hotel right over this blue hole and have tourists cover up your operations, and you would make a killing in tourist dollars while keeping your operation going without the threat of prying eyes? Make this a refuge for the rich and famous. That way, if you get caught, you can list all the important/famous people who stayed here and participated in the festivities."

"Oh, you're good!" said Mr. Rodrigues with bright eyes. "What other ideas do you have?"

"Well," continued Rob. "If you let Leanne and me join your group."

"Ah hem," interrupted Mr. Rodrigues.

"Yes, yes, I know, under strict supervision until you feel you can trust us. Then we can show you how to import your drugs, and all the politicians and many very respected people will be clamoring for your import services."

"Perhaps. Until I make my decision, Carlos, please return our two guests to their room, I mean cell," said Mr. Rodrigues.

POOL TESTING

9 a.m. Monday, July 21st – Pool Test Day 1

“Hi everyone. So how was SeaWorld?” Cody asked.

“Excellent, super, fun.” Diane, Dean, and Thomas responded.

“But that wasn’t the best. The best was the SpaceX launch last night at 1:30 a.m. Now that was amazing,” said Diane as she yawned.

“Oh, damn, I forgot about it,” said Cody.

“Enough on the launch. You can talk about it while we test the AUV. So, everyone, load the cart and head to the pool,” said Brooklyn.

“Cody, you did remember to recharge the batteries last night?”

Cody replied sarcastically. “Yes, Mom, I even turned off the lights.”

“Excellent. Right over there is the pool,” said Brooklyn.

Derek said to Diane. “Don't think it is a big pool, but it will work, and on hot days like this, it is great to get wet.”

The pool was approximately 25 yards long and 10 yards wide, with a maximum depth of 15 feet at one end.

“Luckily, we are primarily using lasers since the acoustics bounce all over the place in a pool like this with flat walls and bottom, which will ricochet the sound waves providing many signals for just one pulse,” said Brooklyn.

They arrived at the pool and reviewed the AUV checklist.

“Power on.”

“Check"

“Wi-Fi connection to the computer?”

“Check.”

“Run the diagnostic routine.”

“Check, running!”

“The lights just came on for 5 seconds.”

“Check.”

“Motor 1 just came on, revved up to maximum speed for 5 seconds.”

“Check.”

“Motor 2, the same.” “Check.”

“Motor 3, the same.” “Check.”

“Motor 4, the same.” “Check.”

“OK, now let’s calibrate the inertial navigation system. Please rotate the vehicle as you would your drone, Cody,” said Brooklyn.

Cody rotated the vehicle around and then flipped the vehicle over.

“Brooklyn, the calibration lights just switched from red to green. All calibrated,” said Cody.

“Excellent Cody. Let’s continue. Tilt the vehicle. Rotate the vehicle to the left, now to the right. Great. The navigation is working fine—next, the lasers. Derek, please check the switch. I am not getting any readings. Oh, here they come. Remind me to replace that switch,” said Brooklyn.

And so, the team checked every component until the checklist was complete.

The last item was to activate the mission to conduct the pool test, which would occur when the external water detection sensor noted that the vehicle was in the water.

“All right everyone. Now, for the moment that we have been waiting for,” Brooklyn began. “If everything goes according to plan. The vehicle will wake up as soon as it enters the water. It will run its wake-up program, where it will immediately learn about its environment. Activate each motor, read every sensor, and save everything to memory. It will be like waking up from a deep sleep and taking note of everything. Dr. Wood, Anthony, Melanie, and I have spent many months writing the code. We are well past 400,000 lines of code. OK, slide it into the water.”

Diane, Cody, and Dean gently lowered it into the pool. A green light turned on.

"The wet switch worked," said Brooklyn. "Now watch."

First, one motor turned on for 5 seconds, then another, until all the motors had been turned on for 5 seconds. Next, various combinations of motors are turned on for 5 seconds. Then, all the motors stopped.

"So, what is it doing?" Diane asked.

"Learning," responded Brooklyn.

"We implemented an artificially intelligent (AI) routine for the vehicle to learn its environment and make difficult decisions."

"Wow," said Derek. "What is it doing now? All the lights and lasers just turned on."

Brooklyn replied that it was activating the system's sensors, such as lights, cameras, lasers, sonar, temperature, and pressure, to obtain the depth and clarity of the water and, in other words, learning as much about itself and the environment as possible.

"Oh look, it stopped," said Derek. "The second green light just came on."

Brooklyn continued. "That means all systems are fine and have a basic understanding of their surroundings."

"Hey, it is moving!" said Thomas.

"Perfect," said Brooklyn. "If it follows what we program, it will follow a star pattern in the pool's center."

Brooklyn sketched a star on a piece of paper for everyone to see.

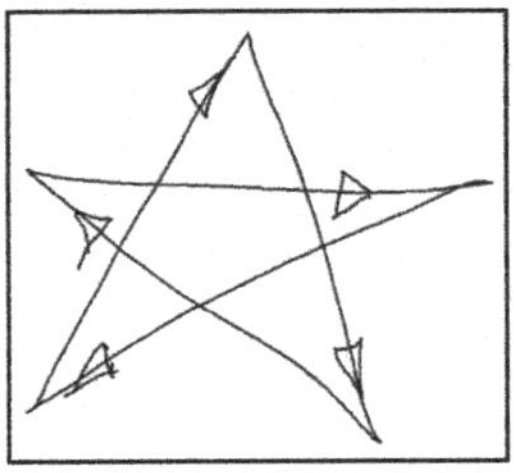

AUV calibration star

"Once that is complete," Brooklyn continued. "The AUV will be ready for its first mission. It will have a basic understanding of itself and the environment stored in memory. Whenever the vehicle turns on, the base configuration will be accessed."

Five minutes later, the vehicle stopped where it first started.

"What now?" Dean asked.

"Now comes the boring part," said Brooklyn. "I need to look at whether the data saved to memory is correct."

"Dean, can you swim to the vehicle and pull it back? Thanks," said Brooklyn.

They pulled the AUV from the water and took everything back to the lab, where Brooklyn connected her laptop to the AUV via Wi-Fi, downloaded the data file, and ran some diagnostic programs to see the vehicle's health.

"Ok everyone. I have many hours ahead of me to analyze this. Why don't you all go to the beach? I will see you all for dinner at the Cove restaurant tomorrow night."

"To the beach! To the beach!" They all cried and marched out the door like the seven dwarfs heading to the mines to hunt for gold.

ψ

7 a.m. Tuesday, July 22nd – Pool Test Day 2

Diane woke up with her phone ringing.

"Hello?"

"Hi Diane, it's Brooklyn. Get everyone up and off to breakfast. I need everyone to be in the lab by 9 a.m. We need to run a series of tests. What we got is perfect. I don't believe it. I want to run every test we ran yesterday and have a few more to add. I will call Cody and Derek."

"Oh, okay. See you at 9 a.m.," said Diane.

She then got on the phone and called Thomas and Dean. An hour later, they were eating breakfast at the Panther dining hall. An hour later, they arrived at the Underwater Technology Laboratory to see that Brooklyn, Cody, and Derek were already there.

"Everyone, let's get the vehicle back to the pool pronto," said Brooklyn.

Like the previous day, they ran all the standard tests. By noon, Brooklyn had started adding other tests. Cody ran for sandwiches and drinks while the others monitored and modified the vehicle software when the slightest deviation from the expected results occurred. By five in the afternoon, they were bushed.

Brooklyn finally called quits at 6 p.m.

"Excellent!! Tonight, I will review the data. Meet me here at 9 a.m. And everyone, get some sleep," said Brooklyn with a smile.

LAKE TESTING

9 a.m. Wednesday, July 23rd

"OK guys, all the data from our pool test were right on," Brooklyn began. "Today, we will take our tests to the next step. The lake we are going to was manufactured. In fact, some of the gravel for Route 95 was mined from the lake. There are a couple of gullies in the lake that need to be mapped. The water is crystal clear, so we should get great photos from the cameras to match the sonar images and calibrate the laser distancing system."

"Are you going to scuba dive it?" Diane asked.

"Yes," Cody replied. "We have our GoPro cameras and will document every vehicle move."

"Can we snorkel in the lake?"

"Yes, go and put on your swimsuits and bring your snorkeling equipment while we load up the Outback," said Brooklyn.

An hour later, they arrived at the lake, where they brought everything to the very large gazebo right on the water.

"Wow, nice lake," said Dean.

"Yes, this is where I trained to become a scientific diver," said Cody.

"Scientific diver?" Diane asked.

"Yes, if you want to do research with a university or government agency, you must be an accredited scientific diver. It takes more than 100 hours of training to become certified," said Brooklyn.

"You will have to tell us more when we get back," said Diane.

They carried the vehicle to the water. The scuba divers suited up with thin diving skins, connected their buoyancy compensators to their scuba tanks, connected their regulators, checked how much air was in their tanks, put on their dive watches BCs with tanks, grabbed the cameras, and slid into the water.

The AUV was then slid into the water, where, 30 seconds later, it came to life just like in the pool. As it came to life, a series of checks were made that took several minutes to complete and then proceeded at a mid-water depth of 15 feet, headed perpendicularly from the shore out into the lake.

The divers followed, filming everything. Everyone else jumped in the water with their snorkel gear and followed the vehicle on the surface.

About 100 feet out, suddenly, the vehicle turned left and dove to 25 feet directly into a gully that ran slightly to the northeast. The gully split after around 150 feet; the AUV followed the right gully until it split; the vehicle took the right gully until it ended.

After pausing for 10 seconds, the AUV turned around and traced its way back until it reached the fork.

Turning right, it followed this gully to the end, where it paused and then followed the gully back to the first branch. It turned and followed that branch until the end.

Once all branches were followed, it traced to where it was put in the water and waited.

Everyone got out of the water.

“Damn, that was perfect! It ran exactly as programmed.” Brooklyn said excitedly to Cody and Diane as they also exited the water. "The vehicle seems to be using all of its sensors, and with its inertial navigation system, it was spot throughout the run!”

Brooklyn continued. “Everyone, it’s time for a food and drink break, and then we will run the same test twice more.”

They returned to the laboratory three hours after completing all the tests.

Diane, who was sitting at the front, asked Brooklyn. “So, what next?”

“I will download everything the vehicle recorded, verify everything, and make mosaic maps with the data: Visual map from the camera photos, acoustic map, and laser map. With these maps, a multimedia movie will be created of the gullies. This will take a few days to complete, so why don't you all take tomorrow off and go to SeaWorld? We will meet the next day to plan our trip to the Spiegel Grove." Replied Brooklyn.

"Brooklyn, I think we should put it in one of the local caves in central Florida first," said Dean.

"Perhaps you are right," said Brooklyn. "Let us plan both. First, we will head to a cave and then to the Spiegel Grove."

ψ

9 a.m. Thursday, July 24th Caves or Shipwrecks

"Caves and the Spiegel Grove?" Cody asked.

"Yes, we will leave at 8 a.m. tomorrow and return in a week. Dr. Wood has rented a van for us. We will be bringing a trailer to carry everything.

"So, where will we go? First, to the Peacock Cave in Wes Smiles Peacock Springs State Park. This cave is well explored and documented so that we can calibrate our vehicle to a real place, but it is challenging."

Diane immediately asked, "Wow, you guys work fast. So, what do we need to do today?"

"Pack up and load up," Cody replied. "Brooklyn has been up almost continuously for the last two days. So, we can't expect her to do much. Pack up the AUV and get it ready for loading onto the trailer. Make sure the second and third sets of batteries are fully charged and ready to be loaded. Next, make sure the paperwork is in order and that we have enough checklists for a minimum of 20 dives."

"Twenty dives!" Said Diane."

"Yes," said Cody. "The data that Brooklyn has compiled show that we designed the vehicle very well and can now have it do some serious mapping and exploration."

"Where was I?" Cody asked. "Make sure the tool kit is fully outfitted, that the dive tanks are loaded along with BCs, regulators, oxygen, first aid kit, dive flag, lift bags, lights, dive pads, floats, a couple of hundred feet of rope, and don't forget the backup dive outfit that Dr. Wood always has ready. Yes, yes, make sure that we have plenty of spare parts in case something fails or malfunctions. Cooler! Make sure the cooler is loaded with plenty of cool packs and filled with water, soft drinks, or whatever else you guys want to drink, and, of course, suntan lotion and bug spray. I am sure that we will think of a million other things as the day progresses."

They all spent the rest of the day getting ready. By 5 p.m., everyone was quite tired. Diane, Thomas, and Dean headed to the dining hall for dinner while Cody and Derek packed the trailer and van.

WES SKILES PEACOCK SPRING STATE PARK

8 a.m. Friday, July 25th – Peacock Spring Underwater Cave

"All right everyone, listen up," said Brooklyn. "The trip to the springs will take about 5 hours with stops. The bathrooms are right over there. Use them. Cody and Derek loaded everything into the trailer last night. Thanks, Cody and Derek. Load your gear. Don't forget your hats, sunglasses, sunscreen, and bug spray."

Diane asked. "Brooklyn, how cold is the water at the spring?"

"Don't worry, Diane." Replied Brooklyn. "You are from up north, so you will be fine. It is a spring, so the water comes out around 55 degrees Fahrenheit, but the surface temperature is usually around 65 to 70 degrees. It is freezing for me, so if I go in, I will be wearing a wet suit."

Fifteen minutes later, they were in the van and heading north. Everyone wanted to know more about where they were going. Brooklyn immediately started telling them what she knew.

"Ok, ok, what I know is that the spring is in the Wes Skiles Peacock Spring State Park. The Park is in a great location with great facilities for cave diving facilities. There are a couple of caves all linked together. We will first send the vehicle down the Orange Grove sink or the Peacock Springs. I would say we start with the Peacock spring. It is the most accessible and popular site in the system. It has an elevated walkway and stairs leading to the spring. Three passageways converge on the spring: the Pothole Tunnel, the Peanut Tunnel, and the Peacock-II Tunnel. Each of the tunnels has a permanent guideline for cave divers to follow that our vehicle can use as a reference. The guidelines start just within the cave system.

"The Pothole Tunnel has a maximum depth of 65 feet (20 meters). It has large open passages with relatively high ceilings. The Peanut tunnel has depths between 20 and 60 feet (6.1 to 18.3 meters) and has crossover tunnels that connect the Peanut tunnel to the Pothole tunnel."

Diane said. "Brooklyn, it says here on my cell phone that the park is 733 acres and is 16 miles southwest of Live Oak, Florida, wherever that is. It also says that it has 33,000 feet of surveyed underwater passages, and it is one of the longest underwater cave systems in the continental United States. It is the 24th longest underwater cave in the world, which took 521 dives to complete by expert divers! No wonder you wanted us to bring so much gear. We will be here for weeks."

Brooklyn responded with a smile. "I wish we could stay for weeks, but those caves have been explored. We will spend a day or two verifying their maps before heading to our next destination."

Derek spoke up. "So where will that be?"

"It's on a need-to-know basis, and you don't need to know yet," said Brooklyn.

"Which means you don't have the foggiest idea or that Dr. Wood still hasn't decided where to send the vehicle for a proper checkout."

At 1:15 p.m., they arrived at the Wes Skiles Peacock Spring State Park, paid the entrance fee, received the key for Dr. Wood's reserved six-person cabin, and talked to the head Ranger about their mission. Luckily, Dr. Wood had paved the way for them to deploy the vehicle. A few years ago, people using diver-assist vehicles damaged the caves, so all vehicles were banned from the cave systems.

They drove into the park, deciding not to waste time in the cabin. They went directly to the Peacock-I spring and found a place to park. Brooklyn told everyone to make a quick survey walk around the area. The guidebook said that the Peacock-I spring was the most accessible and most popular site in the system, with an elevated walkway and stairs leading to the spring. So, that is where they decided to start. As they walked around the park, they saw that indeed, the Pothole, Peanut, and Peacock-II tunnels came together at the spring. One of the divers who climbed out of the spring told them that each of the three tunnels has a permanent guideline or 'gold line' placed there to facilitate the navigation of the cave divers. The guidelines started just within the cave system, so snorkelers or open-water scuba divers cannot see them.

They all returned to the van after 15 minutes of scouting.

"All right everyone, it is almost 2 p.m. We are going to send the AUV from the Peacock-I entrance to the Pothole Sink. As I said earlier, it is 450 feet (140 m) down the tunnel from Peacock-I entrance and is the deepest of the three tunnels, with a maximum depth of 65 feet (20 m). It contains large open passageways and relatively high ceilings, and, according to what one of the divers said as they came out, it has a silt bottom, so we must be careful not to stir it up. They said the walls had been covered in silt but were recently blown away by a big rainstorm last week. Ok everyone, hop to it. Diving in the park ends at 5 p.m., so we don't have much time. Derek, I want you to take the lead today. That way, Cody and I can dive the open portion of the hole," said Brooklyn.

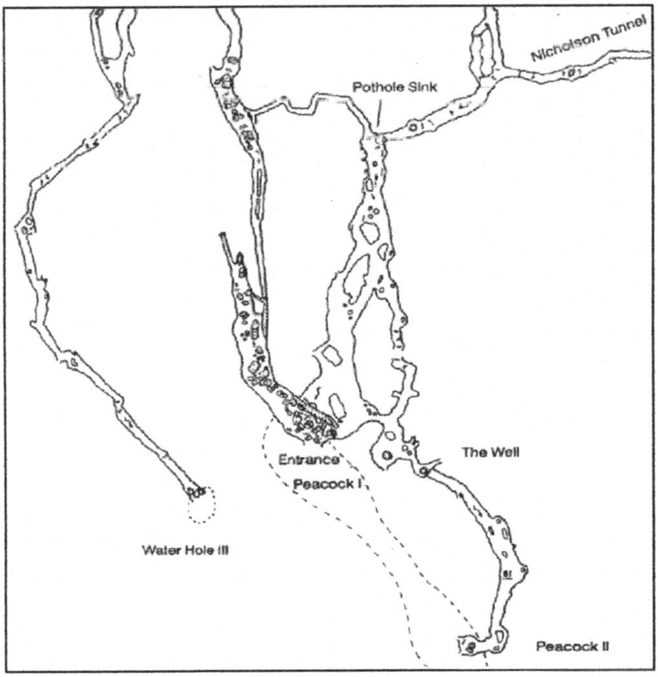

Map of Peacock-I to Pothole Sink

They all jumped to carry everything to the staging area next to the spring. Putting up a small umbrella tent, the team then ran through the vehicle checklist, which they were now very familiar with. Using the maps supplied to them by the park ranger, they entered the basic parameters for the first mission. While Diane, Thomas, Dean, and Derek prepared the vehicle, Brooklyn and Cody got into their wetsuits and scuba diving gear and slipped into the cool water. At 2:30 p.m., the vehicle was lowered into the water, where it immediately activated and ran through a series of tests at the water's surface. Then, it dove and slowly headed toward the cave mouth. Brooklyn and Cody observed it slowly descend and pause at the cave entrance for half a minute and then proceeded into the cave with the mission to proceed to the Pothole Sink 450 feet away and then return. An hour later, it came out of the cave entrance and headed to the surface. Brooklyn and Cody had waited by the entrance the whole time, so they needed to change tanks when they returned to the staging area.

Brooklyn started in: "Now that was one perfect mission. Now, we will send it through the same tunnel and back and then have it continue and map the Peanut Tunnel. As you can see on this tourist map, the Peanut Tunnel is relatively shallow and narrow. They say it is named for a section that resembles the two lobes of a peanut. Its depths range from 20 to 60 feet (6.1 to 18.3 meters)."

Diane piped in: "Just to let everyone know. It is 3:15; we are running out of time."

"Quite right. My Bluetooth link from my laptop computer to the vehicle just went active. I will need about 10 minutes to update the software for the new mission," said Brooklyn.

Fifteen minutes later, the vehicle once again slipped into the water. Brooklyn and Cody, with freshly filled scuba tanks, once again waited at the mouth of the tunnel, watching the vehicle disappear into the depths. Surfacing and waiting at the surface, the two floated about peacefully until fifty minutes was up. Just as they were about to descend, the AUV surfaced and headed to shore, where Diane, Thomas, and Dean lifted it gently and put it on a towel, ready to be rinsed off. Derek helped Brooklyn and Cody out of the water and turned off their scuba air as they took off their buoyancy compensators.

"FYI everybody, it is quarter to 5. Time to wrap it up and head to the cabin," Brooklyn said smiling. "Someone, help me out of this wetsuit!"

They packed up the AUV and carried it back to the van. An hour later, they were in the cabin, sorting out who slept where, and where everything was to go. Brooklyn turned on her laptop computer and began downloading everything the vehicle recorded. Dean and Thomas got out the hamburgers, hot dogs, and veggie burgers, turned on the stove, and started preparing dinner.

Brooklyn took the laser data and sent it through a program she had written to convert the laser data into a three-dimensional image. After twenty minutes, the program ended, saying, 'Processing successful.'

"Hey everyone, who would like to see what the cave looks like?" Brooklyn ran another application, and up popped a view of the tunnel, looking at it from the side. She rotated it 360 degrees and then viewed it from the top.

"Wow! That is amazing!" said Diane.

"Now I am going to take you through the tunnel, and you will see the actual view from the cameras on this screen. Diane, here are virtual reality goggles. Please put it on. You can look anywhere you want. Up, down, left, right, anywhere you want. You will see the real image overlaid with the laser image."

Brooklyn hit the 'Enter' key on her computer, and a matching camera image appeared on one screen exactly at the position of the laser image. The two images were synced.

"Wow! Wow! Amazing!" Diane exclaimed. She looked all over the place. Up, down, all over. This is really amazing, Thomas and Dean; you got to see this."

Thomas and Dean came over as Diane took off the virtual reality goggles. She handed them to Dean, who put them on and almost fell over. "Whoa, that is ... wow! This is the way to view all movies." A couple of minutes later, he handed it over to Thomas.

Over the next two hours, they each put on goggles and were amazed at how they could look anywhere and see everything in amazing detail.

ψ

8 a.m. Saturday, July 26th

As Dean and Thomas piled the pancakes and bacon on the table and poured glasses of orange juice, Brooklyn laid out the plan for the day.

"All right, everyone, we will send the vehicle through the Olsen Sink after breakfast. Olsen is about 1,500 feet (460 m) from Peacock I down the Pothole tunnel. According to the map, the cave system has two small entrances. It says that Olsen was once a popular entrance for cave divers as it allowed easy access to much of the cave and, to prevent erosion, they closed it to cave divers. Derek, you are going to retake the lead this morning."

Brooklyn took a bite of her pancakes, drank some orange juice, and continued: "In the afternoon, we will send it through Orange Grove, northeast of Peacock I. As you saw yesterday, it has a raised walkway and stairs leading to the sinkhole; many people use this entrance. The map shows two winding tunnels that extend from Lower Orange Grove to Upper Orange Grove. The Lower Orange Grove is deeper, extending to 180 feet (55 m). The Upper Orange Grove is a winding tunnel in all three dimensions, starting at 70 feet (21 m) deep and shallows to 40 feet (12 m). This tunnel extends outside of Peacock Spring State Park to the Challenge Sink. So, at 9 a.m. we will start. Finish your breakfast and do whatever else you need to do. Diane, I want you to take the lead in the afternoon. Watch Derek and learn."

Everyone ate their breakfast with gusto and got ready for the day. Suntan lotion, water bottles, soda, snacks, sandwiches in the cooler, batteries checked, etc. At 9 a.m., they all gathered at the van. By 10 a.m., they were at the Olsen hole.

Diane and Thomas immediately started reviewing the vehicle checklist while the others laid out the equipment, spare batteries, and tools. Brooklyn and Cody got into their wetsuits and put on their scuba gear.

"Everything set?" asked Derek.

Brooklyn nodded, as did Cody. Derek then asked Diane, Thomas, and Dean. Everyone responded affirmatively.

Derek said, "OK everyone, Cody and Brooklyn in the water and wait." A minute later. "Dean and Thomas put the vehicle in the water."

Like the previous afternoon, the vehicle went through preliminary tests and then dove. Then, it was a matter of waiting. Brooklyn and Cody came to the surface and climbed out of the hole. They all sat down and had some snacks and a drink. Brooklyn spent the time writing notes. Thomas took photographs, and Derek kept his eyes on the water. The dive was expected to take almost two hours. Two hours and five minutes later, the vehicle came to the surface. As they lifted it onto land, they saw mud stuck to both sides.

"It looks like it found a tight spot, determined it could make it through, and attempted it. Luckily, it made it. I have no idea what it would have cost monetarily and politically if we had to have someone rescue it," said Derek.

"Tonight, I will make a few programming changes to ensure it has a wider area of influence. That way, if the hole is too tight, it will modify or abort its mission," said Brooklyn. "But now it is time for lunch. In 45 minutes, we will move to Lower Orange Grove, and from there, we will have the vehicle explore both Lower and Upper Orange Grove."

By 2:25, they were once again ready to deploy the vehicle.

Diane began: "Now that Dean and Thomas have completed the startup checklist, everything is ready to go. Brooklyn and Cody get into the water and wait." Like that morning, she mimicked Derek. "Dean and Thomas put the vehicle in the water."

Once again, the vehicle was off mapping the cave, and they waited. It surfaced, and they sent it off to explore the Upper Orange Grove again.

At 5:30, they were packing up and heading back to their cabin.

During dinner that night, everyone was excited, and no one could wait to see the footage from the trip. During dinner, Diane asked which cave they would explore tomorrow.

Brooklyn began: "Tomorrow, we will have the vehicle explore the Waterhole Sink. The tourist guide says it contains geological features not seen in the rest of the system, so it should be a good test for the vehicle. Then we will return to the university."

ψ

8 a.m. Sunday, July 27th

When Diane entered the dining room, toast, oatmeal, grits, eggs, and sausages were on the table. "Wow, Dean and Thomas, you two really know how to cook. You both will make fine wives,"

Everyone at the table laughed. Brooklyn said: "As you saw yesterday, the vehicle has functioned beautifully. Just a few more tests this morning, and then off to the open ocean. Let us make this an excellent completion of the cave system. Like yesterday, let's meet in an hour and get everything to the Waterhole.

With smiles, they finished their breakfast and headed out the door. Two hours after that, the vehicle was in the water heading into the Waterhole. The Waterhole was mapped by noon, and they were on the road to the nearest restaurant. That evening, they were once again at the university. As they arrived, Brooklyn began: "I need everyone to unload the vehicle and get it in the lab. I know it has been a long summer, but the real fun is about to begin. So, I want everyone to be at the Underwater Technology Lab at 9 a.m."

"Where are we going?" Diane asked.

"I don't know for sure. I have a hunch, but I can't say anything until Dr. Wood gives the final word. All I know is that it will be fun," said Brooklyn.

By 11 p.m., they were all back at their homes or dorm rooms.

THE SPIEGEL GROVE SHIPWRECK

9 a.m. Monday, July 28th

"Brooklyn, all of you have been talking about the ship, Spiegel Grove. What is so special about it?" Diane asked during lunch.

"You mean that you have not heard of the Spiegel Grove?" Brooklyn asked.

"No, why? Should I?"

"It has a wonderful story. Shall I tell it?" Brooklyn asked.

Diane immediately responded. "Yes, please," With a smile on her face.

"OK, excellent. Let me see. Where should I start? You must know I have forgotten the people's names, so I will fill them in. Now, the story of the sinking of the Spiegel Grove.

"Once upon a time, somewhere off Key Largo in the Florida Keys, a U.S. Navy Landing Ship Dock called the Spiegel Grove was to be sunk as an artificial reef. The ship is 510 feet long and 84 feet wide. It was sunk in 2002 six miles off Key Largo near Dixie Shoals in 130 feet of water. The top of the ship lies at 45 feet.

"As the ship slowly sank, one of the lift bags broke loose and shot over 100 feet into the air. Then another and another until the boat sank out of control and finally came to rest on the bottom upside down. Years passed, and everyone was wondering what to do. Local officials finally concluded that the ship would need to be righted since, as it is, the ship is totally worthless as a tourist attraction. The town council contacted a ship salvage company, which agreed to flip the ship until it was upright. The company sent divers down to the ship and devised a plan to right the ship. A few weeks later, in 2005, they had assembled the ships and equipment. Just as they were about to head out came the weather report – HURRICANE! The hurricane blew through, and after weeks of repairing and rebuilding, the town council restarted the project to re-right the ship. Again, the equipment and ships were prepared, and divers were sent down to review where they were to attach lift bags to right the ship. Thirty minutes later, the divers returned.

"When they were back on deck, the captain asked: 'Hi guys, what is the verdict? Is it a mess down there?'

"'Well, I must say that we have a problem?' David replied, who had spent enough time underwater to be called a fish.

"'What do you mean?' asked the captain.

"'Well, we're not going to be able to right the ship as we thought,' David said with the biggest grin ever.

"'Why? What's up?' asked the captain.

"'The hurricane has righted the ship,' said David.

"'Bull, you are kidding! That ship is 510 feet long and rests at 130 feet deep. No way,' said the captain.

"David repeated, 'The hurricane has righted the ship.'

"'There is no way the ship has been righted. You are making that up! I bet this is one of your practical jokes. I will put up a case of beer every week for an entire year to call your bluff.' Stated the captain before half the crew.

"'I'll take that bet,' said David.

"'Suit up, Bill and Jachec. I want you in the water to verify whether the Spiegel Grove is as David says it is,' said the captain.

"Bill and Jachec were in the water in 20 minutes and back on board in another 30 minutes.

"As Bill came on deck, he called out. 'Captain, I am sorry to say you owe David a lot of beer!'

"That's the story, and it is true, except the names. Can you imagine the power of a hurricane that can flip a 510-foot ship that sits at a depth of 130 ft? That is the story of the Spiegel Grove," said Brooklyn.

ψ

6 a.m., Tuesday, July 29th

"Good morning, good morning! Wake up and get in the van. Cody, check the trailer. Is everything secure?" Brooklyn asked.

"Yes, I double checked this morning. I also checked the lights, tires, and hitch. Everything is good to go." Cody answered with a yawn.

"Everyone, pile your stuff in the car and use the bathrooms; we have a couple of hours' drive," said Brooklyn.

"Great idea," said Derek. "Can we stop for a coffee on our way down? The drive is what - 5 hours. I will need coffee, maybe some of those powder donuts, and a bear claw."

"Yes, yes, we will stop when we switch from Route 95 to the Turnpike on Route 70," said Brooklyn.

Everybody climbed in the van with Brooklyn driving. Everyone was able to have plenty of space to lie down and get some sleep as they headed south. An hour or so later, they were buying coffee and snacks. Back on the road, they talked for a while until everyone, except Brooklyn and Cody, were fast asleep. South on the Turnpike to the Sawgrass Expressway down to Homestead until the highway ended and the access road to the Keys began.

Brooklyn pulled into a gas station. "Brooklyn, it's time for lunch, and I know the perfect place," said Cody.

"Yes, I agree. What do you have in mind?" Brooklyn asked as she stretched to loosen the kinks that built up over the long drive.

"We follow the signs a mile or two toward the Everglades. The little place I know has the best shakes in Florida. I always get their mango shakes, but their passion fruit blended shakes are to die for," said Cody.

"Isn't that the place that has several Galapagos tortoises?" Derek asked.

"Yes, it's a fascinating but very good store," said Cody with a smile.

An hour later…

"Now that's definitely the best shake I have ever had! Cody, you are so right. Okay, everyone. Back in the van," said Brooklyn.

Back on the road again with full bellies and a full tank of gas, they headed to Key Largo.

As they passed John Pennekamp Coral Reef State Park, Diane explained how her family and best friend Bronwyn went snorkeling there last spring to check the condition of the coral.

"There are also several cannons just off the main swimming beach. Next year, after becoming scuba certified, I will return with a measuring tape to document exactly how the cannon lay on the bottom. I don't think what they have on the plaque is correct," said Diane.

"Diane, did you go to Jules?" Brooklyn asked.

"No, what is Jules?"

"The Jules Undersea Lodge is an education center with two underwater habitats. The first underwater habitat has been turned into a hotel. It was called the 'La Chalupa research laboratory' and was used to explore the continental shelf off the coast of Puerto Rico. The second is the Marine Laboratory for Science Education and Research. Dr. Wood said they couldn't put it in deeper water because the giant acrylic sphere that is 6 inches thick has a crack in it. And I mean a very big crack," said Brooklyn.

"Yes, Dr. Wood told us all about it in class," said Cody. "If I remember correctly, he said it cracked due to corrosion!"

"Exactly," said Brooklyn. "The steel at the connection ring where the sphere attached to the habitat rusted. When steel rusts or corrodes, it expands and breaks off into flakes. The corroded area expanded and cracked the six-inch-thick acrylic sphere."

"Wow," said Diane. "That's a lot of force. I can hardly believe it."

"Believe it," said Derek. "I used to tie my surfboards onto my car rack without rinsing the boards. After two years, suddenly, my front windshield broke. The saltwater ran off my surfboards onto the car and down the edges of my windows. The steel rusted, which expanded enough to break the window. Believe it."

"Wow, that is amazing," said Diane,

"Let me continue," said Brooklyn. "If you are a high school student, Marine Lab has a program called the 'Coral Reef Beachwatch Program.' I remember they have a coral reef ecology program in the Marine Lab curriculum. Their Coral Reef Beachwatch program was created for more advanced students to learn about the importance of coral reef habitats. The students get to snorkel multiple reefs and collect coral bleaching data. The Marine Lab is connected to the Mote Marine Lab in Sarasota, Florida, which has created a program for snorkelers to survey the corals while snorkeling. The collected data is entered into an online database used by the scientists at Mote Marine Lab."

"Hey, everyone," continued Brooklyn. "Here is our hotel. Let's check in, drop off our bags, and then head to the boat we have chartered for the next few days."

"Brooklyn, where is the boat?" Cody asked.

"Mile marker 100, on the south side. There is a nice parking lot," replied Brooklyn.

Five minutes later, everyone gathered at the van.

"Everyone, jump in," said Brooklyn.

Fifteen minutes later, they parked in the parking lot with a dive shop, bar, swimwear, and clothing store.

"All right, everyone out. Just around the corner, we should find a store with all types of antique diving equipment and information about the Spiegel Grove," said Brooklyn.

"Hi Brooklyn," Captain George said with a smile as he opened the door. "Come on in. Everyone, grab a seat."

Everyone sat in chairs around a cluttered table.

"So, Brooklyn, what are we doing tomorrow?" Captain George asked.

"Well first, let me introduce you to the team. This is Cody and Derek; they are Sophomores, no, actually Juniors in Ocean Engineering. Here, we have Diane, Thomas, and Dean. They are from a middle school up north, and I can say that they are amazing."

"Nice to meet you all," said the captain.

"Tomorrow, we will load up the boat with our autonomous underwater vehicle. It is totally self-contained. We have gone over all the maps on the Spiegel Grove. We will deploy the vehicle tomorrow around 11 a.m. from the sixth mooring ball attached to the ship near the wheelhouse, which is perfect for accessing and navigating anywhere in the ship. 11 a.m. was selected because there is a slack tide, and when the least current is predicted. The vehicle will descend to the ship clipped to the mooring line in case of a current. Using its video and SONAR, the AUV will proceed to the forward door near the bridge – here," as Brooklyn pointed to the pictures of the ship. The vehicle will proceed at the same level and exit at the same entrance it entered. It will ascend to the surface where we will retrieve it, download the information, and change the batteries," said Brooklyn.

"And what about the divers?" Captain George asked.

Brooklyn was in charge as the PI, also known as the principal investigator. "First, we will need the divers to attach the AUV front clip to the mooring line on the surface. The clip will release when the vehicle arrives at the ship, about 60 feet down. For each run, I want the divers to follow at a distance no closer than fifteen feet unless the vehicle needs assistance. While the AUV is in the ship, it will map the interior using video, acoustics, and LIDAR (lasers). We will then have the vehicle ascend the mooring line so that the vehicle will wait at the mooring line for 2 minutes to allow the divers to connect the AUV clip to the mooring line. When the vehicle is on deck, it is the time for the divers to get out of the water."

"Sounds like you did your homework. Now let us pray that we have calm seas and minimal currents."

SPIEGEL GROVE DIVE DAY

8 a.m. Wednesday, July 30th

"Good morning everyone. This is D-Day 1, as in dive day 1. Safety is the key today. Captain George will give the rundown of the boat, and then I will continue concerning the AUV. Captain George, I hand it over to you," said Brooklyn.

Diane, Thomas, and Dean listened intently as Captain George began. "Hello everyone. First, if there is an emergency, you will find life preservers in these boxes. The Head, or in layman's terms, the toilet, is forward down below on the port side, which is the left side. It is an older model, so don't flush anything other than what comes out of you and toilet paper, and do not use too much, or it will clog. I have a first aid kit and oxygen in the forward compartment. Make sure you follow all instructions from me or the first mate, Alex."

Alex waved his hand and smiled. "I will be handling the dive operations on the boat. Here is Robert, and there is Ted. They are our divers for this expedition. They have been to the Spiegel Grove numerous times, keeping the typical tourist from getting themselves killed. Yes, people have died on the ship."

The captain continued. "It will take an hour to get to the site. There is water and sodas in the cooler over there. Help yourself. If there is an emergency, then this alarm will sound." A very irritating blare came across the boat. "If you hear that, immediately grab a life preserver and come to the rear of the boat here unless you see it is on fire; then we will meet on the boat's bow. I will give you instructions if we have a man-overboard situation. First, yell 'MAN OVERBOARD' if the person is nearby, then throw them a life ring. If not, keep your eyes on him and constantly point in their direction. DO NOT take your eyes off them. In anything but very flat seas, it is very easy to lose track of where they are as soon as they are 100 yards away. Any questions? OK, now sit back, get your vehicle ready, and, of course, make sure you are covered in suntan lotion."

Brooklyn promptly said, "OK, everyone, let's get to work. Like in the lab, Diane, Thomas, and Dean run through the checkout and startup sequence. As you can see, we made sure you have cover from the sun. Without that, you would all be roasting. Anyway, always be sure to stay hydrated with plenty of liquids. This sun and heat can be a killer. Cody, Derek, come over here; I want to re-look over the pictures of the dive site with Robert and Ted."

Everyone assembled at their assigned station, with Robert and Ted on each side of Brooklyn.

Brooklyn started in. "Robert, we will need you and Ted to enter the water first and catch the AUV by its tow clip in the front. If there is any current, you will probably need to pull yourself and the AUV along the Granny-Line to the mooring line. Please attach the front clip of the AUV to the mooring line just below the surface float. The AUV will make its way down the line. The AUV will release the clip when the vehicle arrives at the ship, which should be about 60 feet down. I want you to follow at a distance no closer than fifteen feet for each run unless the vehicle needs assistance. While the AUV is in the ship, it will map the interior using video, acoustics, and lasers. You will know the run is over when the AUV makes a fairly audible beep. You will have one minute to move out of its way. I would suggest going into this room here," as Brooklyn pointed to the map. "The AUV will return using what it believes is the shortest route to the mooring line, where it will wait two minutes for you to re-clip it onto the line. The vehicle will ascend the mooring line on its own to the surface. When you finally surface after whatever safety stops you need, we will need you to assist in bringing the vehicle back to the stern of the ship. When the vehicle is on deck, it is time for you to get out of the water. Any questions or thoughts?"

"Sounds like you all have given this test serious thought. Excellent! Overhead environments such as entering the Spiegel Grove are dangerous, and without special instruction and knowledge, they are not meant for the average diver," Robert said. "We tell divers that there are areas on this ship not intended for access, and for no reason should you be entering an opening below 90 feet or the main deck on the Spiegel Grove, even with advanced certifications. It is good to hear that these tests are in areas that are easily accessible. If, for any reason, you need us back on the surface, tell Alex. He has a special signaling device. Tell him A) Emergency - We need to return ASAP with the vehicle, B) Emergency - We need to return ASAP without the vehicle, or C) Cancelling test, return with vehicle."

"Perfect," replied Brooklyn. "OK, everyone, let's prepare for a long day."

For the next hour, they worked steadily, getting everything ready. The captain spotted the mooring lines and aimed for the number 6 mooring line. Luckily, the only boat already on the Spiegel Grove was moored to the last line, the number 8 mooring. The captain radioed the other boat and informed them of the tests we will be performing. They replied that they would be out of the water in two hours and not in any position to get in the way. And wished them success with the tests and that they would love to see them if we got some excellent results.

The open line attached to mooring-float number 6 was grappled by the boat hook and attached to the bow cleat.

"What a morning!" thought Brooklyn. Perfect weather, no waves, and can you believe it – almost no current; there must be nearly 100–foot visibility. Wasn't that a hint of the ship down there? "Captain, is that the ship there?" She asked.

Everyone rushed to the side to look.

"Yes, indeed it is. You guys have hit the jackpot," said the captain.

"Everyone to the stern – the rear of the boat – let's get the divers in the water and then the vehicle," said Brooklyn.

At 11 a.m., precisely, the AUV started descending the mooring line. Occasionally, the divers thought the vehicle was getting stuck. Still, they watched as the vehicle angled upward each time the horizontal thrusters started, loosening the connection and allowing the AUV to continue its downward journey. At the bottom, the vehicle waited. The divers approached and pushed the release button on the clip. Ten seconds after the release with lights on, the sonar pinging, and the lasers scanning, the AUV rotated around with a swish and headed directly to a point near the opening selected for penetration. The sensors that detected the vehicle's velocity (the DVL or Doppler Velocity Logger) noted a slight crosscurrent and corrected the vehicle's orientation to account for the drift.

Coral Growth on the Spiegel Grove

"Perfect!" thought Robert. The vehicle arrived at its starting location only ten feet from its starting point. The vehicle immediately detected the entrance and entered at a very slow speed. The video cameras, sonar, and lasers began an in-depth mapping of the deck. When the vehicle located an opening to an inner room, it rotated and entered. Keeping track of where it was and what was on every side, sometimes it moved vertically up and down or left and right to map areas that were hard to map. One of the rooms took over 10 minutes to map since it was filled with all types of debris. At the opposite end of the ship's deck, the AUV stopped and made a loud audible beep. The divers moved into the side room, watched the AUV proceed, and headed down the corridor toward where it came in. The divers followed at a distance and only just caught up to it when it arrived at the mooring line. Ted grabbed the mooring line, and AUV snapped the clip tight and moved away. Soon, the vehicle was seen moving up. Robert and Ted proceeded cautiously up the mooring line, taking their necessary safety stops, glad they were using rebreathers for the dive.

George came to Brooklyn once the AUV was on the deck and the divers were on board. "Shall we head back?"

"Yes, and once we are underway, we can discuss tomorrow's mission," said Brooklyn.

The mooring line was released, the divers changed out of their dive suits, and everyone gathered around Brooklyn and the captain.

"So, what is on for tomorrow, Brooklyn?" asked the captain.

"Tomorrow, we will leave the dock at 9 a.m. and proceed to the same mooring buoy exactly as today. We will send the AUV down to 80 feet and have the AUV go to the second deck of the main superstructure. We will start with the port side aft (rear) entrance. The corridors on this level go in a square around the entire deck, with many small rooms on both sides for the vehicle to explore. Hopefully, the vehicle will not get snagged on the rope guideline that runs down one section. The divers will always have visible natural light, and access holes have been cut into most rooms. In some of the rooms, there are tables, cookers, cabinets, bathrooms, and toilet cubicles. From what I have read and what people have told me, there is very little silt on this deck and plenty of space to access each room. The AUV will exit by the same opening it entered by determining the fastest route it believes it can take and by following that route out. I have programmed the vehicle to ascend the mooring line again. Like yesterday, the vehicle will wait at the mooring line for two minutes to allow divers to connect the AUV to the mooring line. Any questions?" Brooklyn asked.

"If everything goes as well as it did today, we will get you all the data and the proof that your vehicle is fully operational. From what Ted and Robert said, the AUV ran the track without incident. It caught on the line in the one corridor, but how you initiated a backup and change in the distance from the side wall worked perfectly," said Captain George.

An hour later, they approached the dock. What a successful day they had. On the pier, a boy about Diane's, Dean's, and Thomas' age waited patiently to grab the bow line. Suddenly, Diane shouted with excitement.

"Hi Zack! What are you doing here? It is great to see you! How did you know that we were here?"

"Hi Diane," said Zack as he cleated the line to the dock. "Through the scuttlebutt."

"The what?" cut in Diane.

"The scuttlebutt. The scuttlebutt is the rumor mill where rumors spread throughout the keys. It came from the old sailing term. I was cleaning an old guy's boat on Marathon Key when I overheard two sailors talking about a team Dr. Wood sent down to test an autonomous underwater vehicle. When they mentioned that several middle school students would be involved, I figured you would be one of them," said Zack.

"How did you figure that?" Diane asked.

"Because that's just who you are. You are the type who gets into all types of really cool adventures without even trying," said Zack.

Diane just beamed. "It's so nice to see you again." And gave him a big hug as soon as she jumped off the boat. The two walked up the dock with his arm over her shoulders as she told him about the AUV and what it was for.

Dean nudged Thomas in the ribs. "I didn't know Diane had a boyfriend."

"It sure looks that way," replied Thomas with a grin. "We are never going let her live this one down."

"From the look of it, I don't think she cares," said Dean.

DEAD BODY!

10 a.m. Thursday, July 30th

"Another perfect day. Perfectly calm seas, almost glass," said Diane.

"George, how does the afternoon look? Any thunderstorms predicted?" Brooklyn asked.

"Yes, by 4 p.m., we better be back in the harbor. Until 2 p.m., it should be fine, but the winds will pick up heavily as the thunderstorms build. When we get there, no dawdling. It will be the same tomorrow if you need another day," said Captain George.

"Everyone, gather around; today will be the same as yesterday, except that the vehicle is going down to the next deck below. The entrance is about 80 feet deep. I looked over some of the results from yesterday's dive. Very, very nice!" exclaimed Brooklyn with a big grin. "I hope today will be as good. Dr. Wood will be ecstatic about the results that we are getting. Robert and Ted, are there any changes at your end from being so deep for so long?"

"No, but it will take longer for us to come up. You may want someone to swim up the Granny Line and grab the vehicle. I wouldn't want it to bounce around too long if the seas start to build up," said Ted.

"I can do that," said Zack. Diane had asked Brooklyn late the previous night if Zack could come along. She relented after much discussion and realization that he could be a wonderful asset.

"OK," said Brooklyn, "As soon as we arrive, we need the divers to get in the water as soon as possible and, as yesterday, immediately get the vehicle in the water."

At 10:45 a.m., the vehicle descended to the ship. Like the previous day, the divers videoed and took photos of the AUV as it made its way to the lower deck. Arriving in front of the entrance, the AUV halted and spent a minute scanning the area. Slowly entering the ship, it began mapping and documenting everything. In the third room, it entered and scanned all the debris. Suddenly, it stopped and repeatedly rotated - back and forth - around something resting on the deck. The divers couldn't see what the vehicle was analyzing from their angle. They noticed that a red LED light turned on the back side of the vehicle. After another five minutes, it proceeded out the door. Both divers entered the room to see what it was. There, lying in a grotesque position, was a body. They came very close and saw the reality of the scene. Someone had cleverly placed a manikin to look like a dead body. Nodding to each other, the divers followed the AUV on its mission of mapping the ship. Soon, the AUV came to the end of its planned run, where the AUV halted and once again sent out a loud audible beep. The divers once again moved into the side room and watched the AUV go by, heading back to the opening it came in by. The divers following caught up with the vehicle as it arrived at the mooring line. Robert grabbed the mooring line, tightened its front clip, and moved away. Robert and Ted came on deck a little while later to find everyone huddled around the AUV, looking at the light on the back of the vehicle.

George came up to Brooklyn and asked, "What is it?"

"The warning light is lit. It says that it found something seriously wrong during its travels," said Brooklyn.

"What constitutes something serious?" asked the captain.

"Well, at the moment, only three things are programmed. 1) There is a hazardous substance that should be avoided at all costs – but we have not implemented the sensors for that; 2) There is a very dangerous situation, and divers should not enter – but we have not quite figured out what it should do if it finds itself in this situation; and 3) It found a dead body. So, that leaves only one thing. It found a dead body in the ship using its onboard picture analysis software," said Brooklyn.

"Oh, no! Let's find out what the divers saw," said the captain.

Soon, the divers were on board. "Robert, Ted, the vehicle returned with an emergency light on. What happened down there?" asked the captain.

"You will never believe it; the AUV found what it thought was a dead body. It scanned it for ten minutes. Once it finally moved on, we entered the room and discovered a professional-looking manikin lying in a grotesque position. It was very lifelike. If it weren't for the little details of how a dead body looks underwater, you would take it for the real thing," said Robert.

"I won't ask you how you know that," said the captain.

"Guys," said the captain to the AUV team. "It looks like your vehicle found a hoax. The body it found isn't real. So, when you review the data, realize it is not a real body."

"Whew," said Diane.

"Wow," said Brooklyn. "I think we have some really interesting data in this vehicle. Robert, you said the vehicle scanned the manikin for over ten minutes and was very lifelike?"

"Yes," responded Robert, and Ted nodded his head.

"Everyone, I think we can head home now. From what the divers said, the vehicle handled itself perfectly, and I need to review all the data. That alone should take a week. And if you all hadn't realized, it is time for Diane, Thomas, and Dean to return to their homes.

ψ

10 a.m. Friday, August 1st – Everyone piled into the van and soon headed north. They could return to the university by early afternoon, bypassing Miami's traffic jams. For the next hour, they unloaded the van and moved everything back into the lab. Brooklyn returned the trailer to its parking spot and returned the van to security. It was 4 p.m. when all tasks were completed.

"Diane, Thomas, Dean, Dr. Wood, and I will be in your dorm at 11 to pick you up and drive you to the airport. So, you have tomorrow to yourselves. Cody or Derek, if either of you has time tomorrow, perhaps you can take them somewhere," said Brooklyn.

Cody stepped forward. "No problem, the dance tonight was canceled. Something about having to redo the wood floors: I know just the thing for these adventurers. We are searching for silver, gold, gems, and jewels. I will pick up Dr. Wood's underwater metal detector this evening. Diane, Thomas, and Dean be in front of your dorm at 6 a.m. and make sure you have already covered yourselves with suntan lotion."

"Are you joking, Cody? Silver, gold, gems, and jewels," said Diane.

"Joking! Never," said Cody with a smile. "Perhaps you don't know your Spanish history about the Queen Jewels. They even made a movie about it called 'Fools Gold.' In the early seventeen-hundreds, the King of Spain was to marry and had to provide a dowry to the new queen, so he ordered all types of gems and jewels for the queen to be made. Well, the fleet with all the treasure made it to just off the east coast of Florida when a hurricane hit them. All but one of the ships ran aground and sank between Ft. Pierce and Cape Canaveral. Every month of the year, silver, gold, gems, and jewels wash up on the beaches. In fact, if I remember correctly, one of the ships sank right off Melbourne Beach, the Santisima Trinidad y Nuestra Señora de la Concepción. Dr. Wood is assisting a team in conducting an archaeological excavation of the ship. He often documents the site, mapping, recovering, and conserving artifacts.

Underwater Crawler

He supplied an underwater crawler to support the dig. If we find anything in the water, we will have to turn it over to the archaeological team, but if we find anything on the beach, it is ours. Are you guys interested, or will you waste a day suntanning at the pool?"

"Really?" Thomas asked excitedly.

"We are going with you," said Derek.

"This will be a sweet ending to a fantastic trip!" said Diane.

"Perfect. See you all tomorrow. Bright and early. Oh, and don't forget to bring your snorkel gear. I want to check out an anchor that a buddy of mine spotted while he was lobstering. It is in about ten feet of water about a hundred yards out just north of Wabasso beach." Cody walked out the door and headed for his car.

SEARCH FOR TREASURE

Dawn Saturday, August 2nd

"Good morning, Diane, Thomas, and Dean. Get in the car. We will first visit a few sites just north of Vero Beach that often provide gems and jewels. Then, we will go to Wabasso Beach to find the anchor. The pictures he took of the anchor showed that it was from the early seventeen-hundreds," said Cody.

"Hey, what is all this stuff in your SUV?" Diane asked.

"That is the metal detector there. These are sieves to filter out the sand to find small items. These are here for digging and filtering out big stuff like rocks and things, and these long-handled strainers are actually for fishermen to collect 'sea fleas' that live just under the sand at the waterline in Florida. They are also excellent in searching for treasure, and here is my fat-wheeled buggy, 'Little Red,' to carry everything," said Cody.

They arrived at a beach just north of Vero Beach an hour later. They unloaded the SUV. Walking through the sea-grape-covered canopy pulling 'little red,' they came to a pristine beach with beautiful turquoise water. Small, one-foot waves gently washed up the beach. The sun was still low on the horizon, and white puffy clouds spread across the sky.

"Everyone, see that pine tree over there? That is where we need to go. The stern of one of the shipwrecks is right off the beach from there. Diane, grab the metal detector. While we haul all this stuff to the site, I want you to check the water's edge. It is almost low tide right now, and the best opportunity to check this is close to the water. Remember, don't go into the water. That is out of bounds for us. Every inch of the water from Ft. Pierce to Port Canaveral is permitted, and I don't know who has this permit," said Cody.

As Diane pulled the metal detector from the pile of stuff on 'Little Red,' Thomas asked. "What do you mean permitted?"

“I mean that various people have permits issued by the State of Florida to have the exclusive right to search this area for treasure, which means that if you are caught going into the water with a metal detector, you could potentially be arrested. It is like poaching.”

“Really?” asked Thomas. “I thought the oceans were for everybody.”

“Ha. Think again. If there is something of value anywhere, there is always someone who wants to lay claim to it. In general, only a few people are searching this area. The key point in laying claim to a shipwreck is to make sure it is not military or has never been associated with the military.”

“Why is that?” Dean asked, who was huffing and puffing, pulling ‘little red.’”

“Well, let’s put it this way. The United States has a great number of sunken military ships and submarines. So, Spain and UNESCO brought up the subject in a lawsuit to recover the gold found in a Spanish galleon that an American company in international water recovered. The United States jumped at the opportunity to make it an international law to keep everyone’s hands off their sunken wrecks. It is complicated because the ship did not have to be in the Navy when it sank. It must have been in the Navy at some point during its life. For example, if the Navy took a mail ship for four years during World War II and sank 20 years later in a storm, it would still be classified as a sunken military wreck. If the mail boat was delivering something of value and a salvage company wanted to retrieve it – nope. They will not even be allowed to touch or explore the wreck without Federal Government permission.”

“Wow! That is crazy,” said Thomas.

“Yup, so just stay out of the water with the metal detector,” said Cody.

“Cody, how do I work this thing?” Diane asked as she puzzled over the device.

“Easy. First, turn this nob here. That turns it on. The further you turn the knob, the louder the speakers are. Keep it somewhere in the middle; as for the other buttons, turn this one to automatic and leave the others where they are. Here, let me bury a quarter in the sand.”

Cody buried the quarter about eight inches down. "Now pass the detection ring over the area like this." Cody stretched his arm out in a comfortable position and swept the area with the metal detector's ring about an inch over the sand. As it passed over where the quarter was, a loud sound came from the headphones of the metal detector. "Oh, that is too loud. Let me turn it down a bit. OK Diane, your try."

Diane imitated Cody in stance and form and swept the area with the detector. Immediately, the headphones made a tone.

"Thanks, Cody! Now for the search."

"Keep everything you find, especially anything you cannot identify," said Cody.

Cody, Dean, and Thomas made their way to the beach in front of the large pine tree.

"Dean and Thomas, could you help me put up this canopy? When the sun gets higher in the sky, it will be scorching," Cody said.

Fifteen minutes later, the canopy was up, 'Little Red' was unloaded, and a blanket was laid out. Diane showed everyone what she had found: 1 soda can, a sand-worn earring, two bottle tops, and a screwdriver.

"Excellent start," said Cody. "Grab a drink from the cooler. Two of you go back and forth one hundred yards down and one hundred yards up the beach from here. Try to cover everything from the waterline to the sand dunes. Check the waterline first. I need one of you to come with me to scout the water area. Whoever it is, grab your mask, fins, and snorkel, and meet me in the water."

Diane, Thomas, and Dean did rock scissors paper until it was decided that Dean and Diane would do the metal detecting on the beach, and Thomas would go with Cody.

Diane and Dean went off, sweeping the detector from side to side. When Diane detected something, Dean would dig it up and try to identify it. Everything big or anything immediately recognized as trash went into a garbage bag, and everything small went into a gallon-sized clear plastic bag.

Cody and Thomas headed out into the water.

"Thomas, I want you to stay fifteen feet to my right. Always keep that distance from me no matter where I go," said Thomas.

"What are we looking for? Thomas asked.

"Anything unusual; look for patterns; look for straight lines. Anything from 1715 that isn't buried will be completely encrusted with coral or other living creatures. Often, something of interest will be hidden in plain sight. Don't expect to find anything, but be ready to spot anything. If you find something, click this clicker."

Cody passed Thomas a small plastic device that made a loud clicking sound when pressed. Thomas passed his hand through the loop attached to the clicker and tightened the strap so he couldn't lose it.

The two proceeded out and swam perpendicular to the beach for about 100 yards. Cody turned north, swam 30 feet, and then returned, swimming toward the shore. Out and in, they swam. Each time, going further north. Suddenly, Cody heard the loud click of Thomas' clicker on their fifth transect (that is what they are doing, swimming transects).

"Thomas! What have you found?" Cody asked.

"I don't know, but it is something man-made. Come over and look."

Beam from a shipwreck

Cody swam over while taking out his small underwater GoPro camera. He looked down and dove to the bottom, which was only ten feet deep. He swam over the area back and forth, taking multiple pictures from different angles each time. Three times, he surfaced for air before swimming back to Thomas.

"Excellent, Thomas! It has nothing to do with the 1715 wrecks, but it certainly is from some wreck. There are hundreds of wrecks throughout Florida waters. I took several photos, but I don't think we will find much to identify it. The one thing we need to do is anywhere there is sand around the structure, dig gently into the sand. Sometimes, little things like pieces-of-eight or gold doubloons might accumulate just under or next to large objects such as stones or objects like the one you found. Let me tie this dive marker to the structure, so I have nothing in my hands. Then let's swim down and check it out."

A minute later, Cody was back. "OK, let's do it!"

Cody and Thomas dove down and swam along the edges. It was encrusted almost everywhere with coral and thousands of other living critters. Thomas found a corner with sand. Nothing. Cody saw another piece with some sand and dug around. Nothing. They came to the surface and dove down again. In a nearby indentation in the coral, Thomas found a pocket of sand. Immediately, he dug with his fingers. Something? What did he feel? Is that a coin? Holding his breath longer than ever before, he pulled out a coin. Zooming to the surface, he gasped for air and yelled to Cody. "I found a coin! I found a coin!"

Cody zoomed over and said, let me see it. Thomas, who hadn't actually looked at it yet, handed it over to Cody.

"Very nice, Thomas. You just found a 1983 quarter. It has been in the water for at least a year. Great find. See, I told you you could find coins in those locations."

Nothing else was found after an hour of swimming. Everyone met back at the canopy, laid down, and had a sandwich and a cold drink from the cooler.

"I would like to spend one more hour here before heading to our next location. Diane, come with me," said Cody. "Thomas and Dean finish what has not been searched and then come in and have a relaxing swim before we head back to the car."

Dean asked. "What shall we do with all the garbage we have collected? Boy, there is a lot of it."

"We will take it with us and throw it out in one of the bins at the parking lot," said Cody.

Fifteen minutes later, they parked in the Wabasso Beach parking lot. "Now, over there is a great sandwich shop. I'm going to get a bite. These guys make the best sandwiches and have all kinds of curiosities you can buy. Diane, you were talking about getting some souvenirs and gifts. This is a good place to pick them up," said Cody.

Thirty minutes later, with full bellies and a bag of gifts and souvenirs, they again started to load the SUV and load 'Little Red.'

"OK, this is a small trek. You see that house there, the one about 15 houses down. That is where we are going. One of my friends found a large anchor about 100 yards from that house. The anchor looks like it is from one of the 1715 Spanish Plate Fleet ships. We are going to find it again," said Cody.

"Now that's cool. We see cool things like that on TV! I never thought we would actually see something truly historic," said Diane. "The cannon I saw in Pennekamp Park on Key Largo were reproductions. This is so cool!"

Twenty hot minutes later, they put up the canopy, laid out the blanket, and got ready to snorkel out to the site. Cody got them all together on the blanket under the canopy. "All right, I don't know the exact spot, so we will break into two teams. Team one will be Diane and me. Team 2 will be Thomas and Dean. First, we will swim directly to a spot that I selected. I will put this dive marker at that location. That will be our starting point. We will then make a circle search pattern. Team 1 will go clockwise, and Team 2 will go counterclockwise. The visibility is excellent today, so stay spread out. I would say have fifteen feet between you. Once we meet up, we will go out another 30 feet from there and circle back the same way we came. The person on the outer loop on the first run will be the inner person on the reverse run. That person needs to keep the edge of what he saw coming out in view as the team circles back to the starting point. Once we finish one loop, I will move the marker sixty feet to the north. We will do this three times and then move sixty feet further out into the ocean and head back south. Any questions?"

No one had questions. They reapplied their 100-factor suntan lotion and headed to the water with their snorkel equipment. Swimming easily in the warm water, both teams were soon in position. Cody blew a whistle, and everyone started swimming about fifteen feet between each person in their assigned direction. Dean from team two stopped suddenly and yelled, "Shark! I see a shark!"

"Stay right there! Don't move, and don't panic. I will be right there," yelled Cody.

Cody swam as fast as he could and quickly arrived where Dean was floating.

"Where did you see it, and what did it look like," asked Cody.

"Right there," pointed Dean. "It looked like it was 10 feet long. It was resting on the bottom!" replied Dean nervously.

Cody looked down to where Dean had pointed. Nothing was there. He rotated all around until he had almost completed a circle when he saw it.

"Don't worry, Dean. You saw a nurse shark. You can still see it there." Cody pointed in the direction he had seen it. "They are fairly harmless. You may lose a finger or two if you feed it dead fish. If you leave it alone, it will leave you alone."

Cody then yelled to the others. "Guys, swim over here and see a nurse shark."

Diane and Thomas swam over and could dive within 15 feet of the shark as it lay lazily on the bottom. Ten minutes later, they were once again searching for the anchor or anything that might look like it came from a shipwreck.

They completed the first area and were on the way to the second area when Diane shouted excitedly, "There it is! Cody, it is right there! I see the anchor."

Everyone swam over to where Diane was and looked down. Sure enough, there it was, partially buried and very big. It must have been 15 or maybe 20 feet long, with one of the arms and most of the shank visible but heavily encrusted with all types of stuff growing on it.

"Wow, that is a big anchor. It must have been from one of the ships carrying the 'Queens Jewels,'" said Cody.

They spent the rest of the day diving around the anchor, hoping to find a piece of eight or a gold doubloon. As the afternoon turned toward evening, they returned to their car and the university for dinner and to pack for their flight home the next day.

ψ

9 a.m. Sunday, August 3rd – It was a sad day for Diane, Thomas, and Dean. It had been an exciting month with thousands of new impressions, hundreds of new experiences, and a whole new set of friends. Now, they had to fly home. Yes, they still had about a month of summer vacation to go, but the adventures they had in this month would last a lifetime. Advanced underwater vehicles, caves, shipwrecks, sea turtles, NASA, Sea World ... Wow!

As Dr. Wood drove them to the airport, they quietly looked out of the windows and remembered the weeks of adventures they just had. As Dr. Wood said goodbye to each other as they passed through security, he also said, "See you soon. You never know what comes up and where people's lives will lead. So, I expect our paths will pass again quite soon. Have a safe flight."

A few hours later, their parents greeted them with open arms. Boy, did they have stories to tell. Yes, that had kept everyone up to date with what they were doing, but it was different from talking on the phone versus talking directly. The stories continued into the evening when exhausted parents finally sent them to bed.

MISSING!

10 a.m. Monday, August 11th

Dr. Wood came rushing into Brooklyn's office a week after the team returned from a successful Spiegel Grove trip.

"Brooklyn, we must get the AUV to the Bahamas today!" Dr. Wood looked agitated.

Brooklyn promptly asked, "Dr. Wood, what has happened?"

"Two friends were cave diving on Abaco Island of the Bahamas. They didn't check in when they were supposed to. Their boat was found at the entrance of one of the many blue-hole cave systems on the island," said Dr. Wood.

"Oh no! What can we do?" Brooklyn asked.

"I want the AUV to explore the cave, find my friends, and bring them out," said Dr. Wood.

Thinking for a moment, Brooklyn said: "Dr. Wood, there is no one here. Cody and Derek went on a European tour and will not return until the beginning of the semester."

"What about the junior high kids? They were really bright," asked Dr. Wood.

"I can ask, but I know Dean was heading with his family to Yosemite for their annual family reunion. I will call Diane and Thomas to see if they have passports available."

"Yes, passports, I forgot that. You have a passport, right?" Dr. Wood asked.

"Oh, yes. If you remember, I was at the Oceans Conference in Asia last year."

"Excellent. Give Diane and Thomas a call. Is there any way we can deploy it without them?" asked Dr. Wood.

"Yes, but there are so many variables that the risk of missing something is high. Also, we need the manpower to lift and deploy it," said Brooklyn.

An hour later, Brooklyn was on the phone with Diane.

"Diane, we have an emergency here. We need to bring the team together! Are you available to come back? I mean to fly out tomorrow morning?" Brooklyn asked.

"Wow, yes, I mean no, I mean, I have to ask my dad. What happened?"

Brooklyn quickly explained to Diane what had happened.

"Brooklyn, my mom, and dad must be on this call. Can we Zoom in an hour? It would be best if Dr. Wood were with us on the Zoom," said Diane.

"Yes, I agree. See you in an hour," replied Brooklyn.

An hour later, the connection was established, and after introductions, Dr. Wood filled Sam and Suzanne in on the situation.

"… so, it comes down to this: we need the AUV to explore the cave system to hopefully find out what happened to Rob and Leanne. Unfortunately, my team has dispersed for the next month. Only Brooklyn is around. We wondered if Diane, Dean, and Thomas could return to deploy the vehicle. What do you think?" Dr. Wood asked.

Suzanne started first. "First of all, if there is any danger, we are not going! I will not let my daughter come close to danger!"

Dr. Wood jumped in immediately. "No, definitely not. This is only a rescue mission or perhaps a search-and-find mission."

Sam, Diane's dad, piped in. "Dr. Wood, what are the chances they are still alive?"

"Well, not high, but those caves are known to have chambers with air pockets that have ledges and open areas. We hope that they were able to get to one of these. There is too much territory for divers to cover the entire cave system. It would also take too much time. Especially since we are approaching peak hurricane season," replied Dr. Wood.

"OK," said Sam. "Let's make this happen. Diane, you said that Thomas and Dean are not available. What about your friend in the keys? Zack somebody. What about him?"

"I am sure he is available. When we talked at Pennekamp Park, he said he was going to cement coral sprigs to ceramic discs. A very boring job when there are 10,000 coral sprigs to glue."

"Excellent. Give him a call," said Dr. Wood, "tell him that I am covering the costs. In fact, the same goes for you, Sam, and Diane; just let me know who, when, what airline, etc. Keep your receipts. Take the first flight out. Fly directly to Melbourne, Florida, if you can. I will set up the flight from here to Abaco Island. Plan to stay on the island for a week. I cannot see how we will need more than that."

After a few minutes of discussion, they ended the Zoom call.

Sam began, looking at his wife, "OK, who goes with Diane? We are not letting her go alone, but with little Ben and Freddie, one of us must stay here. I have a big presentation on Thursday that I can postpone a week."

Suzanne said thoughtfully. "I am pretty much open this week except for the PTA meeting on Tuesday. But I really think this is for you and Diane to do. You two are the mechatronic geeks who love to play with robots."

"OK, I'll see what I can do to reschedule my presentation. Diane, get packed," said Sam.

"I want to go, too! I want to go, too!" Ben said multiple times.

"Sorry Ben," said Sam. "This trip is not for you. We will Zoom and call every day. We will have our GPS linked on our phones, so you will know exactly where we are at all times?"

"Daddy, I just talked to Zack. He will be on the next Greyhound bus heading north to Melbourne. He will be at the airport when we arrive," said Diane excitedly.

"Perfect now to set up Flights," replied Sam.

FLIGHT TO THE BAHAMAS

Noon, Wednesday, August 13th

“Hi Zack!” As Diane rushed over to hug him.

"Hi Diane. Dr. Wood will be here in a few minutes with the van. Brooklyn is already at the plane loading the AUV and other supplies."

Sam promptly said, “OK, that gives us a little time. Everyone over to the restaurant; we have supplies for the next day or two.”

Twenty minutes later, with bags in their hands, while munching on sandwiches and drinking their sodas, Dr. Wood drove up. “Hi everyone. Great to see you! Our flight takes off in an hour on the other side of the runway. I will drive everyone there.”

“Hi Dr. Wood,” said Sam." I brought my scuba gear in case we needed to dive. I am not cave-certified, though."

“Me neither,” said Dr. Wood. “I also brought mine.”

Another twenty minutes later, they loaded the two-engine Beechcraft Super King Air with all their stuff and soon climbed onto the plane. Dr. Wood and Sam sat in the first row, Diane and Zack in the second, and Brooklyn in the third row.

"Wow,” said Diane as the wheels left the ground. I have never been in a small plane before. Zack, have you?”

“Yes, and in helicopters. I was with my dad, who works for the Florida Keys National Park. We flew across the Keys documenting the damage from Irene.”

“Irene? Who's Irene?” Diane asked.

"Hurricane Irene," said Zack seriously. “It was really bad, and the reefs, as you saw last winter, suffered a lot of damage.”

“I remember now. By the way, where is your dad?”

“He is on the research ship WeatherBird from the Florida Institute of Oceanography, verifying that the barge that ran aground on Molasses Reef left an oil slick on almost every island. That company will be hit with a very, very big fine,” said Zack proudly.

"He let you go without questions?" Diane asked.

“Oh, he asked many questions, but he knows Dr. Wood by reputation, and he understands the importance of it, and when I told him who else would be on the rescue mission, he immediately purchased the bus ticket and gave me enough money for the trip,” said Zack.

“Wow, your dad is cool,” said Diane.

“Sometimes, but he has a real issue concerning grades. I can do almost anything I want if I am a straight ‘A’ student, but if I go to B or C, the hammer falls, and the fun stops. So, I study my butt off and keep the straight A's, which allows me to go on adventures when school is out,” said Zack with a smile.

An hour or so later, the plane touched down and headed to the small terminal at Treasure Kay Airport. At the front, after passing through customs/border police, they found a small bus with seating for six and a lot of space for luggage/equipment, which they filled full to the top.

Sam noticed that the bus was not normal.

It was more of an expedition bus with all types of extras like a generator with gas mounted on the roof, awnings mounted to the roof rack that would open to both sides of the bus, an extra air-conditioner also mounted on the roof, a sink, a showerhead, stove, porta-potty, and so on.

“Where did you find this?” Sam asked Dr. Wood.

“I have had it for a couple of years. When I heard about Rob and Leanne, I drove it to Fort Lauderdale and had it shipped here. It has most of what anyone would need, including satellite and radio antennas. It is perfect as a mobile operating base or MOB. I call this MOB-1.”

Diane, who was listening, immediately asked. “So, what is MOB-2?"

"That would be my boat," responded Dr. Wood. "Unfortunately, the boat is being serviced," said Dr. Wood.

"What he means," Brooklyn said, "is that his 65-foot ketch is being fitted with all types of gadgets and cannot go to sea for another week."

"OK, everybody, we will stay at the Treasure Cay Hotel Resort & Marina tonight. Tomorrow, we will drive as close to the site as we can. If we cannot get there by bus, I have reserved a boat to take us there," Dr. Wood said.

They arrived at the resort a few minutes later, and Dr. Wood went to check-in.

"I have booked a suite for the week. It has three bedrooms. Brooklyn and Diane, you are together. Sam, if you are okay with Zack, I will take the smallest room by myself," said Dr. Wood.

"Sounds like a plan. Everyone, get your bags," said Sam.

They all piled out of the bus with their things and went to the suite. An hour later, they met in the large room. Dr. Wood had two maps open. The first was of the island, and the other was of the cave system that Rob and Leanne had entered.

Dr. Wood began, "Here is what we have and what we know," pointing to a place on the map. "This is where we know that they entered the caves. The police found a few of their things at the entrance."

Everyone looked at where he was pointing. "The police say we can get the bus to right here." He pointed to a place slightly south of the entrance point. "Tomorrow, we will leave here at dawn, make our way to that point, and set up a base camp. Sam and I will make an orientation dive to see what is at the entrance to the cave. At that point, we will make further decisions about deploying the AUV."

They all pondered what he said for a minute.

Dr. Wood continued: "Diane, Brooklyn, and Zack, you know more about the AUV than anyone worldwide. You guys are professionals. Sam and I will take any advice you can give. Don't ever be shy. It would be best if you spoke your mind. OK?"

All three nodded.

“This afternoon and evening, I would like you three to unpack the AUV and run through your tests. Make sure all the batteries are charged and that everything works perfectly. Any questions?” asked Dr. Wood.

“Nope, okay, we will meet for dinner at 7 pm. Sam, we have some recon to do.”

Sam replied. “Do I need anything special?”

“No, first, we are going to Cooper’s Town to talk to various storekeepers, taxi drivers, and the like. You see, there is something fishy here. Rob and Leanne are just too good at what they do. They would never get into trouble doing a cave dive or any other adventure. I find their level of caution irritating. So, no, if they are dead, then they were murdered. My gut feeling is that they found something they shouldn't have and are now hostages,” said Dr. Wood.

“O... K,” said Sam slowly, pondering what Dr. Wood had just said.

Dr. Wood continued. "It has something to do with the cave or the cave system they entered. I am sure the locals will inform us of the details.”

"In that case, I will grab my shoulder bag with a few useful things in it,” said Sam.

Starting with the shopkeepers, Sam and Dr. Wood asked about the non-traditional clientele, people who shopped there, and whether anyone stood out. From there, they went to the one taxi company in the area, then to a couple of the local harbors, where they talked to the harbormaster and the gas attendants who refueled the boats. The best sources were two old guys passing the time away in front of the bait shop and an old fuddy-dud with only a couple of teeth hanging out in front of the only bar in town. At 6:30 p.m., Dr. Wood said to Sam, “That was very informative. Let's head back and try to put everything we learned this afternoon together in some logical order. This island seems to be where things are happening—Chinese, Russians, CIA, NSA, Mexicans, Cubans, Colombians, Japanese tourists, etc. I am sure some of this is innocent enough, but the ones that stand out are the Chinese and the Columbians. It was interesting that Rob and Leanne asked some of the same questions. Those two old guys at the bait shop. Boy, do they hate the Chinese! Of course, don't blame them if they actually did the damage to their fishing and to their island that they say they did."

"Yes, but as you said, we will be in the area where the Chinese company was and will be able to access the damage personally," said Sam.

At 7:30 p.m., they were all sitting around the dinner table at the Spinnaker discussing everything they learned. They sat overlooking the ocean long after the food was gone.

"Well," said Sam. "It seems that the people who might be responsible would-be Colombians or Mexicans. I don't think the people here know the difference."

"If it's the Columbians, it's probably drugs. But, as they say in town, they rarely see any of them. I bet that if they have an import/export base somewhere around here, they supply it at night by seaplane. We are so close to Miami and Fort Lauderdale that it would be easy to run a food and supply operation once a week without anyone getting suspicious," said Dr. Wood thoughtfully.

"OK," continued Dr. Wood. "What do we know?"

Diane chimed in, "The Chinese company started building a shipping port on the island's north side, but when the United States found out, they convinced the Bohemian government to shut it down."

"The Russians tried to get involved, as did the Japanese, but to no avail, which accounts for them, but all this happened a year or two ago. I wonder why we never heard this on the news?" said Brooklyn.

"Probably not to scare people. You may remember reading in your history books about the Cuban missile crisis. Many people were terrified," said Sam.

"Yes, and Cuba is the same distance from the United States as this island is," said Dr. Wood.

"That leaves the Cubans, or is it the Mexicans or the Colombians? Why would a Cuban care about this island? They have their island near the United States," said Zack.

"So, does anyone have a different possibility of who or why someone would be here? Do we all agree that it is probably all about drugs," said Dr. Wood.

Everyone nodded.

"So, how do you think they are smuggling drugs?" Sam asked.

"Good question," said Brooklyn, "planes, trains, boats, submarines, cars, balloons, what other means of transportation are there?"

"Well, balloons, trains, and automobiles are out," said Zack.

"That leaves planes, boats, and submarines," said Brooklyn.

"Planes and boats are the obvious choices, but I am sure they would have been caught if that is the case. That leaves submarines," said Sam. "How are they doing it? How are they getting the drugs into the US using submarines? Where did they get the sub?" Sam asked.

Dr. Wood thought and thought until he smiled; a grin covered his face. "Brooklyn, remember the story I told when we were all at Ichabod's across from the anchorage? It was about a couple of friends of mine, Luc, Arun, Gustavo, and Humberto, back when I was doing my master's at the University of Miami. We had all gone scuba diving off Fowey Rocks Lighthouse south of Key Biscayne. When we returned, the border patrol stopped us and asked where we were from. I said USA, Luc said France, Arun said India, Humberto said Columbia, and Gustavo said Argentina.

"The customs agent said, 'You've got to be kidding.' We had just named all the primary drug countries in the world. They searched the boat up and down, in and out. Of course, they found nothing, but I noticed they never checked the empty scuba tanks. They didn't even open a value. I thought to myself, what a way to smuggle in drugs?"

"Yes, I remember the story," said Brooklyn. "You often tell it."

"Hmm," said Dr Wood. "A couple of years ago, I was stuck in Key West for four days until the WeatherBird Research Ship returned.

It was the night when I had five British sailors bowing down to me, saying that I was their God, but that is another story. That night, I had too much to drink and told the story to the British sailors. There was this Hispanic guy nearby with some cute lass on his lap. He attentively listened to every word I said. An hour later, he bought me a drink and asked me who I was. I laughed and gave him my card. He said his name was Rodrigues and went on his way. I have not thought of that moment since then. I wonder if he took my story and put it into practice?"

"Could be," said Sam, "but I want to hear that other story."

"What other story?" Dr. Wood asked.

"The one about the British sailors," replied Sam.

"Yes, yes," said Diane and Zack.

Brooklyn nodded her head.

Dr. Wood looked at them all. "OK, let's see. It was a weekend of gay pride celebrations. There I was, ordering a beer at the bar, when five guys with bald or very closely cut hair barged their way through—almost knocking my beer out of my hand. I thought, great, more rowdy Gay Pride celebrators. Well, the guy who knocked me over, turned and apologized. I overheard what they were talking about when I realized they were British sailors off a destroyer. The destroyer was to conduct hurricane studies and was expected to go into every hurricane that came by that year. Of course, that was the year we didn't have any. So, we started a conversation. I told them who I was and what I did. These guys were my guys! Two operated the autonomous and remotely operated vehicles on the ship. The other three were responsible for underwater demolition and removing unexploded ammunition. There was fantastic conversation until this cute blonde came in. They all had eyes for her. Little did they know I had met her with her husband the day before. They were from Canada for a friend's wedding. The night before, I had danced some swing dancing with her. Her husband doesn't dance, and she loves to. All the sailors were focused on her as a dog on a fresh bone. Each asked her to dance, but none knew how to swing dance, so she became bored with them. So, I strode forward and asked her to dance. She recognized me but didn't let on. We took to the floor and did a whole set of really wild moves. Her smile was a mile wide. When the song ended, she hugged me and went to where her husband was sitting, unknown to us. At that moment, all the sailors started bowing down to me, saying that I was their god and asking how I learned to dance like that, especially since I was a professor. Professors are fuddy-duds who don't dance. Well, that is that; I was scuba diving off the Dry Tortugas the next day. I never heard of any of them. I wonder what they are doing?"

"Wow," said Sam. "That is quite a story. Whew!"

"Ok, everyone, to bed; we have an early start tomorrow," said Dr. Wood.

BLUEHOLE

6 a.m. Thursday, August 14th

Everyone was bleary-eyed as they ate breakfast. Just as they finished, a policeman and a guy who looked like a public official came to their table.

"Good morning, Mr. Pintard. Excellent, you were able to come. Anything we need to know?" Dr. Wood asked.

“Hello Dr. Wood. I see you brought your team. I didn't know your students were so young,” said Mr. Pintard as he looked at Diane and Zack.

“Oh, don't look at their ages. These two will make your mouth drop at what they are capable of,” pointing to Diane and Zack, “This team is one of the best in the world. I would not trade any of them for the best engineer out there.”

Mr. Pintard smiled with a big smile. “I will take your word for it. I have a policeman, Sergeant Mitchell, and one of my junior administrators, Mr. Thomas, to assist you and smooth the way if someone wishes to stop or hinder your search.”

“Thank you very much. We are happy to accept any assistance,” said Dr. Wood.

“Keep me updated. I wish you the best in searching for your friends,” said Mr. Pintard. Mr. Pintard walked out.

The policeman and the administrator came forward.

“Hello, I am Sergeant Mitchell, and this is Mr. Thomas; we are here to assist. Also, if you need anything, contact me if it concerns the law. If it has to do with policy, talk to Mr. Thomas.”

“Nice to meet you both. Let me introduce my team," said Dr. Wood. Pointing to each in turn. “This is Sam; he is Diane's dad and a mechanical engineer. This is his daughter, Diane, one of our experts with the autonomous underwater vehicle. Then we have Brooklyn, my graduate student, the team leader, and Zack, who is great with everything marine."

"Nice to meet you. When you are ready, let us know. We will be waiting in the Land Rover," said Sergeant Mitchell.

"As Sergeant Mitchell and Mr. Thomas went out the door, Sam said. "What a nice reception. I would almost guess they are here to watch rather than help us. Only time will tell what the truth is."

"OK, time to hit the road. Load up the bus," said Dr. Wood.

The AUV was lifted into the bus first. All the batteries were fully charged. The vehicle had been thoroughly checked the day before. Just as they were about to close the door, a pickup truck quickly stopped beside them. A woman in her mid-thirties jumped out of the vehicle.

"Svetlana. How wonderful to see you!" exclaimed Dr. Wood.

Everyone was amazed at Dr. Wood's greeting as he hugged her.

"Everyone! Meet Svetlana Novikova; she was my first graduate student. She now runs one of the dive shops on the island when she isn't gallivanting around the world. So, Svetlana, what is new with the deep-sea magnesium mining in the Pacific?"

"Dr. Vood, it is lucky ve vere able to protect the best areas before the piranha arrived. I vas able to dive ten times on Voods Hole Oceanographic Institute's submersible Alvin and three times on Russia's Mir-I submersible to map and document the protected areas. But enough on me. Show me the vehicle," said Svetlana eagerly.

"Brooklyn, show her the AUV," said Dr. Wood.

Brooklyn ran through the AUV's specifications, always wanting to show off what she had helped invent. Svetlana asked question after question, with Brooklyn answering most of the questions, but Diane answered many.

"Dr. Vood, you always have top-notch helpers, but I didn't know you had superb kids on your team. I am really impressed. How old are you, Diane?"

“I am eleven,” answered Diane, a bit taken aback.

“Vonderful, Dr. Vood, your new vehicle is amazing. I vish I could come with you today. Perhaps tomorrow? Vould that be okay?” Svetlana asked.

“Positively,” answered Dr. Wood. “Do you have the scuba tanks and supplies?"

“Yes, in the truck. Six aluminum 80 tanks and a cooler full of sandviches, snacks, and drinks,” said Svetlana with a wink.

“You are the best! See you tomorrow,” said Dr. Wood.

Diane and Brooklyn both said, “Bye.”

They all jumped onto the bus with Dr. Wood driving. The main road curved to the left, and just a short drive down a dirt road went north, where they turned to come front to front with a line of gigantic concrete blocks blocking the way.

“Now, this is a pain in the behind," said Dr. Wood. “It really looks like someone doesn't want vehicles down here.”

“So, what are we going to do?" Diane asked.

"Walk and carry everything to the Blue Hole," replied Dr. Wood. “But first, let's go and check it out. We need someone to watch the cars. Any volunteers?”

No one raised their hands.

“OK, it looks like we will have to go to the tried-and-true method of selection: Rock-paper-scissors. Ok, everyone, pair up.”

“Dr. Wood,” said Zack, “I'll volunteer.”

"Really! That is good of you. Thank you. I'll remember it. Here are the keys to the bus. Here is a walkie-talkie. I wonder if we have reliable cell phone signals so far out. Food is in the cooler. I don't think we will be long. Everyone, grab a bottle of water and some snacks.”

They all headed off on foot, with Sergeant Mitchell and Mr. Thomas following along in the rear. Ten minutes later, they started to see signs of development and construction. The dirt road was very rough from the heavy construction vehicles that had passed that way. A broken-down backhoe was visible through the trees.

"Looks like the company left in a hurry," said Sam. "It looks like they left everything they couldn't easily remove!"

They rounded a bend and found several mobile homes and a scaffolding tower in poor condition—one without a door. Looking briefly inside, they were empty except for a chair and a rusty metal desk.

"From what the locals say, this used to be a beautiful place. They say to go north from the huge Banyan tree to find the Blue Hole. Well, there isn't one. It looks like the bastards killed a beautiful tree and left it a mess. No wonder the locals hate them," said Dr. Wood.

"Over here. I think I found what is left of the tree. This means the Blue Hole should be north of here," said Sam.

Up and over gravel mounts and rough terrain, they came across the Blue Hole with massive cuts on one side as if someone used a gigantic chainsaw to cut the sides; the cuts were very deep, perhaps fifty to sixty feet deep. The water was a deep, deep blue and very clear. Over to the left, less than a hundred feet away, they could see the inlet and the beautiful turquoise sea beyond.

"Now that is beautiful!" said Diane.

"Oh, yes! And what a mess this place is," said Brooklyn with disgust.

"I suggest we come back by boat. This trek will kill us, and it won't do the AUV any good either," said Dr. Wood as he almost toppled over. Sweating from head to toe, he looked very bedraggled.

"Let's check out the inlet and the path we will have to take from there," continued Dr. Wood as he wiped his face with a towel. "I don't think the hotel will want to clean that when you bring it back," laughed Sam.

"Nope, I doubt it. That is why I will leave it in Brooklyn's and Diane's room."

"No, you're not!" exclaimed Brooklyn as Dr. Wood and Sam burst out laughing.

Except for the cuts made with heavy equipment, the inlet was beautiful.

“Time to go back to the bus and get the AUV. You will find 'Little Red' on the bus’s roof,” said Dr. Wood.

“What! You have 'Little Red?' That will make it a lot easier,” said Brooklyn, who looked like she had climbed a mountain.

Back they trudged to the bus, where they found Zack sitting on a lounge chair in the shade of the bus, drinking an iced tea.

“Any problems, Zack?” Dr. Wood asked.

"Nope, and I don't think we will,” said Zack.

“Why not?”

“Two things, not one car went by since you left, and the police car. I think that may be a great deterrent,” said Zack.

“You are probably right. Well, get off your duff, get on the roof, and get 'Little Red' down,” said Dr. Wood.

“What’s 'Little Red'?" Zack asked.

“The little red wagon with big fat wheels that we carry the AUV with,” replied Diane.

“Oh,” said Zack as he climbed the ladder attached to the bus’s back door. Ten minutes later, they loaded the AUV onto the cushioned bed of ‘Little Red.’ Piling whatever they could onto the buggy using bungy-straps to hold everything in place.

Sam grabbed the front handle, and Zack pushed from behind. Brooklyn and Diane carried specially designed backpacks that held the delicate electronics that might be needed. Dr. Wood pulled a two-wheeled cart with big sand tires carrying a small generator and some spare batteries for the AUV.

Back to the Blue Hole, they trudged. It was approaching noon when they arrived. First, they set up the folding canopy and put the AUV in the shade; they then took a half-hour break for lunch.

“Hey everyone! These sandwiches are fantastic! What did your friend put in them, Dr. Wood?” Zack asked.

Everyone quickly grabbed one and took a bite. Mms and aahs were heard all around,

"Now we have everyone together. Brooklyn, take charge here and get the AUV ready for deployment. Sam and I will take 'Little Red' and get our scuba gear. Any questions?" Dr. Wood asked. "No? Ok Sam, let's go." They headed back to the bus.

Brooklyn immediately started. "OK, time to prep the AUV. We have spare parts, batteries, and tools if we need them."

Diane, Brooklyn, and Zack got the preparation procedure sheets and ran through the familiar checklist. An hour later, as they were finishing, Dr. Wood and Sam rolled the cart full of scuba equipment and tanks.

"Whew, that is a workout," said Sam.

"Yes, I am definitely not going to the gym this week," said Dr. Wood.

"But you told us that you never go to the gym. You always say that if one keeps active doing cool stuff, you never need to work out," said Brooklyn.

"So right you are. Zack, you are familiar with scuba, right?" Dr. Wood asked although he knew the answer. "So, give us a hand getting the dive gear ready."

A half-hour later, Dr. Wood and Sam were geared up and ready to go. The AUV was next to enter the water.

With her notepad out, Brooklyn asked, "Dr. Wood, how many psi?"

Dr. Wood and Sam each had an 80 cubic-foot aluminum tank rated 3000 pounds per square inch of pressure (psi).

Looking at his pressure gauge, Dr. Wood promptly said: "3,050 psi." Sam immediately said, "3,075 psi."

Dr. Wood continued, "We will not be long. We will check the cave entrance and return in about fifteen minutes."

"Ok," said Brooklyn as Dr. Wood and Sam sank below the surface. Brooklyn wrote in the notebook and said aloud: "12:45 p.m. Dr.

Wood and Sam submerged." Then she continued: "OK, we have a few final checks to the AUV."

Five minutes later, Diane said, "That's it, now to put it in the water and let her go."

"Everyone, take a seat; we have a few minutes," said Brooklyn.

Twelve minutes later, Dr. Wood and Sam surfaced.

"1:02 divers up," Brooklyn said as she wrote the same in her notebook.

"Well," said Dr. Wood, "everything looks normal. Rob and Leanne's guide rope is there, and runs off into the cave. I went in about 50 feet, nothing."

"What is the clarity of the water? Were there any currents?" Brooklyn asked.

"One-hundred-foot visibility or more, and I didn't feel any currents. Sam, what is your assessment?"

"I agree; it is the clearest water I have ever been in, and no, I also didn't feel any currents."

"Brooklyn, let's get the vehicle in the water. Sam and I will guide it to the starting point. You are using the line tracking algorithm at first, right?" Dr. Wood asked.

"Yes, followed by the general search and map algorithm that Melanie and Anthony wrote. It worked perfectly in the Peacock Springs caves a month ago," Brooklyn responded with a smile of confidence.

"Excellent. Time to get it wet," said Dr. Wood,

Brooklyn, Diane, and Zack lowered it into the water, with Dr. Wood and Sam assisting from the water. Thirty seconds after the water sensing sensor detected it was in the water, all its systems turned on. Using the knowledge of all known maps of the cave system and its current GPS position, it slowly set off toward the cave opening; Dr. Wood and Sam followed it down to the entrance. The AUV paused and rotated left and right, mapping the entrance and comparing it with the maps it had in memory. When the maps needed to be corrected, they were updated with new three-dimensional cave maps. Slowly, it entered the cave, swinging left and right periodically. At this point in the search, the line laid by Rob and Leanne was its primary navigation focus. Off it went, out of sight. Dr. Wood and Sam surfaced and exited the water. Everyone made themselves comfortable. Sergeant Mitchell and Mr. Thomas approached Dr. Wood and started a conversation.

"Dr. Wood, what is that machine?" asked the Sergeant.

"Sarge, think of it as an autonomous search system. It can search for hours and map every inch of where she goes. There is a laser mapping system, acoustic mapping system, video, and still photos. She has a robotic arm in case she needs to grip something and wire-cutters if she gets stuck on some fishing line. We have included several tools for many situations normally found in a shipwreck or cave," said Dr. Wood.

"How much does it cost?" asked the Sergeant.

"Now, that is a good question. I really don't know. You see, in actual parts, it costs less than $10,000, but in man-hours for the designing, building, testing, and programming the vehicle. Who knows? I am sure that if some organization decides to mass-market the vehicle, it will be around $100,000 to $200,000 since my students have done all the hard work."

"Dr. Wood," Mr. Thomas started with a bit of hesitancy. This was really out of his comfort zone. "How long do you think it will take for today's search?"

"Well, Brooklyn and Diane have told it to search to the end of the line that my friends Rob and Leanne left in their dive. If it finds anything like a dead body, to come right back."

"It can find dead bodies?" Thomas asked with amazement in his voice.

"Yes, it found one in the Florida Keys deep in a shipwreck," said Dr. Wood.

"The mission today will take a maximum of four hours. After it checks the line, it is to work its way back, searching all side passages one at a time until all have been searched. If any of these is very long, it is to explore a maximum of one hundred yards and then return, documenting that exploration is incomplete. If the time limit is reached, it will return directly. So, sit back and relax; we have at least an hour, maybe four, before it returns."

With that, Dr. Wood laid back on the blanket he had brought along, covered his eyes with the hat he had on, and napped. Sergeant

Blue Hole Site

Mitchell and Mr. Thomas found somewhat comfortable places to wait. Brooklyn, Diane, and Zack swam about in the beautiful water for almost an hour before they too, laid back and waited. It was closing at 5 p.m., and the vehicle still had not appeared. Dr. Wood sat next to the water's edge for twenty minutes. Concern was seen on his face. Brooklyn was with him. Diane heard them talk about something that sounded like a transponder and an underwater modem.

"Brooklyn, can we set up the vehicle's transponder to act as an underwater modem?" Dr. Wood asked,

"Yes. It's already installed, but it uses too much electrical energy to keep it running," said Brooklyn.

"I understand that, but can you reconfigure it so it turns on for five minutes every hour to send us updates? Right now, we only have it broadcast if the vehicle is in trouble," said Dr. Wood

"Yes, I can make the changes tonight, but we should reduce the mission time. It is set for a maximum of 8 hours with an operation time of 4 hours. I will change it to a maximum time of 6 hours with an operation time of 3 and a half hours, which the vehicle can modify to 4 hours by the AI (artificial intelligence) program that Melanie and Anthony had programmed," replied Brooklyn.

At twenty minutes past 5 p.m., the AUV emerged.

Zack was the first to spot it as it came up from the depths. "There it is! Boy, it looks like it has been in a fight."

There were scrapes on its sides and mud caked to the bottom.

"It looks like she ran through some tight spots," said Diane as she wiped some mud off.

Brooklyn took photos of everything for documentation. Dr. Wood immediately said, "OK, let's get the little lady onto the cart and back to the bus. We will need to make two runs with 'Little Red."

Without further discussion, they loaded the AUV onto 'Little Red,' piled whatever else they could onto the cart, and headed back to the bus. By the time everything was loaded on the bus, it was dark. Sergeant Mitchell and Mr. Thomas were about to get in their Land Rover when Dr. Wood spoke up.

"Sergeant Mitchell and Mr. Thomas, we will do the same thing tomorrow. See you tomorrow morning?" Dr. Wood asked.

"Yes, see you tomorrow," said Sergeant Mitchell.

ψ

"First things first. Dinner! I know I am starving, and I am sure you all are. Back to the resort, and let's get some food," said Dr. Wood.

"Brooklyn and Diane, do you know why the AUV took so long?" Sam asked.

"I think it got stuck," answered Brooklyn.

"Yes, definitely," Diane continued more confidently. "All the scrapes and the mud prove it."

Brooklyn continued. "We will know everything when we download the data."

An hour later, with full bellies, they transported the AUV from the bus to their room, where Brooklyn immediately started downloading the day's data, almost one terabyte of information. At the same time, Diane and Zack set up the vehicle's batteries for recharging. No one said anything. Sam and Dr. Wood watched the team in amazement.

Sam said to Dr. Wood: "Boy, they work like a well-oiled machine. I have never seen the like with kids so young."

"Don't tell them that, or it will go to their heads, and we will never hear the end of it," said Dr. Wood.

It was nearly 9 p.m. when Brooklyn said, "All right, everyone. The data are downloaded and matched by the vehicle's clock. I can now run everything fast forward. As you can see, there are four monitors. This one to the far right is the SONAR or sound images. This one to the left is for the LIDAR or laser images. The monitor to the right is the normal video with the compass heading, depth, tilt, and velocity overlaid onto the monitor so we can readily see what is happening. The forward view is the biggest front-and-centered since six video cameras are on board. The right and left cameras are here and there," said Brooklyn as she pointed out the camera views. "Here and here are the top and bottom cameras, and this is the view from the rear video camera. Over here, to the far right, are all the other data we don't care about now, like salinity, turbidity, oxygen, radioactivity, PH level, and so on."

"Nice," said Sam.

"Brooklyn and Diane, please run the video at twice the speed," said Dr. Wood.

Brooklyn increased the speed of projection. Everyone's eyes were on the video monitor since the water clarity was fantastic. After watching the scenes zoom by at twice the speed for 5 minutes, Dr. Wood had them up to four times the speed.

"Yes, let's keep it at this speed for a while. Everyone, keep your eyes peeled for anything out of the ordinary. Yell if you see anything. This will take about an hour at this speed.

It was about an hour later when Diane yelled, "I thought I saw something

Brooklyn slowly reversed the sequence of data and video presentation. The vehicle was at the end of Rob and Leanne's guidelines. Something was at the edge of the visible light on the cave floor. Something that looked like a strap.

"Anyone recognize that? It looks like a strap to me," said Dr. Wood.

The others nodded while Dr. Wood continued, "Well, we have a good reason to return tomorrow. Brooklyn, take an image of that scene and print it out."

Brooklyn copied the best view of the strap and printed it.

"Now, let's go through the rest of the data," said Dr. Wood.

They had gone through the rest of the video at four times the speed by 11 p.m. when the AUV turned itself off and was pulled from the water. There was nothing new of an exotic nature, but they had great data. The cave now was 3-D laser mapped, 3-D Acoustic mapped, and 360-degree video mapped such that, when viewed with virtual reality goggles, you feel that you are the vehicle flying through the cave. You can see what is around the vehicle no matter where you look.

"Time for everyone to go to bed; we have another long day tomorrow," said Dr. Wood.

POSSIBILITIES

7 a.m. Friday, August 15th

"Everybody, time to wake up. Breakfast in 20 minutes. Wake up." called Dr. Wood.

Groggily, they all crawled out of bed, muscles aching from all the strenuous work they had done the day before. Thirty minutes later, they were at breakfast, and as they were finishing up, Sergeant Mitchell walked up with a more rugged-looking guy approached.

"Hi Sergeant Mitchell," said Dr. Wood, looking from Sergeant Mitchell to the new guy, "and who is this?"

"Hi Dr. Wood, this is Mr. King. He is replacing Mr. Thomas."

The way he said that showed what he really thought of Mr. Thomas.

Mr. King was a huge guy with the biggest smile and the whitest teeth. Diane looked at his hands. They were huge!

Dr. Wood looked and looked at Mr. King. "Have we met Mr. King? You look as old as I am, perhaps a few years younger. You look familiar."

Mr. King laughed a big hearty laugh. His skin was darker than dark; his teeth were amazingly white, and his smile would win over the most skeptical.

"Think back, Dr. Wood. Miami, South Tower dorms, twelfth floor, parties, Calypso. Now, do you remember?"

"Arthur! Wow! What are you doing here?" Dr. Wood asked excitedly. "Boy, it has been a long time. I never expected to meet you again!"

"Me neither. I moved here soon after we graduated from the University of Miami. I met my wife during our last semester. She is from Abaco Island. When I came here to meet her family, I fell in love with the place."

"Excellent, so how did you get this job chaperoning us around?" Dr. Wood asked.

"Volunteered for it. I got a call from Mr. Thomas complaining about the heat and bugs and how boring it was chaperoning a batch of children around. Well, I know he is a sniveling little snot, and I was wondering if it was you," said Arthur King.

"Do you remember Rob, who was also on our floor in Miami?" Dr. Wood asked

"Yes, the guy from Saskatchewan, Canada; I remember him," said Arthur.

"Rob and his girlfriend Leanne were cave diving here and have gone missing," said Dr. Wood.

"Oh no! I heard this was a search and rescue mission or, more likely, a search and find mission, but I never expected to know the people. What can we do?" Arthur asked.

"Help us in the search. We are heading to the site right now. We sent our autonomous underwater vehicle to the cave where Rob and Leanne were known to be diving. At the end of our search yesterday, we saw this." Dr. Wood handed Arthur the picture of the strap. "This strap should not be there, and it was recently dropped or left there."

"Yes, I see."

"Today, we will focus on starting the search from that point. The vehicle will go directly to where the strap is and continue the search," said Dr. Wood.

Diane came up. "Dr. Wood, Brooklyn, and I have an idea. While the AUV continues the search, why don't we mark on the surface where the cave system is? We have all the navigational data from yesterday's search. Using a compass, it will be easy to mark where the cave passages are on the surface."

"Great idea," said Dr. Wood

"Everyone, this is Arthur King, a very good old friend. He will be helping us with the search. Let's get going," said Dr. Wood.

A hotel staff member handed Dr. Wood a note as they got up. Dr. Wood read it and said to the crew, "Svetlana will be meeting us at the site. She will have her boat to support the search."

Zack immediately said: "I hope she is bringing lunch! Those were the best sandwiches ever!"

They all left the breakfast room and headed to their suite. Soon, they once again had the bus loaded up. It went quickly since Arthur could pick up the AUV by himself without any effort.

Back to the site, they drove. During the dive, the music playing was interrupted by the radio announcer, who said someone called Dorian was on his way.

"Hey, turn it up," said Arthur. "We need to keep track of Dorian."

"Who is Dorian?" Diane asked.

"Not who, what! Dorian is a hurricane formed in the middle of the Atlantic. Now, the meteorologists have no idea which way it will go. In six hours, they hope the hurricane models will be able to tell us more," said Arthur.

They parked precisely as they had the day before but immediately got down 'Little Red' and loaded it up. Within a half-hour, they started at the Blue Hole. An hour later, as they were preparing the AUV, a sixty-foot trimaran slowly entered the bay, inching its way closer until it finally dropped its anchor. A woman was seen rowing the boat's rowboat to two other locations 30 degrees off the stern to drop two more anchors. The boat was now positioned with the stern close to the beach.

Everything was fixed on the ship and positioned to support the AUV team. Soon, the rowboat was on its way to the shore,

"Hello Svetlana; great to see you could make it," said Dr. Wood.

"Hi Dr. Wood, Brooklyn, Sam, Diane, and... Arthur, what are you doing here?" Svetlana asked.

"Hi Svetlana. That's not the right question since I am here on a government directive. What are you doing here? I thought you were in Albania on an underwater archaeological dive," said Arthur.

Everyone was watching with high interest. What an exciting thing to be doing. Diving in the waters where ancient Romans, Greeks, Phoenicians, and Egyptians sailed and finding and bringing up ancient artifacts that no one had seen in thousands of years.

"The archaeological expedition was shut down for the year. It got too dangerous with so many people trying to steal it. First, we had diving bandits; then there was the typical thief in the night; then there was this company who tried to stop us because they said they had the right to dig the site and sell everything. They were in cahoots with some politicians. Ultimately, the military stepped in and shut down the whole project until the mess could be resolved. So, I came back here to make some money taking tourists out," said Svetlana.

"Ms. Svetlana," began Diane.

"Just call me Svetlana."

Diane continued, "Excuse me, but what happened to your accent?"

Svetlana laughed. "I use that to throw off new people I meet. People always expect me to have a heavy accent because of the name Svetlana. It confuses them when I talk to them without a Ukrainian accent," replied Svetlana.

Brooklyn came up. "The AUV is ready to launch. Were you going to dive today?"

"Yes. Excellent. Sam, let's get suited up," said Dr. Wood.

Fifteen minutes later, they were sliding into the water. Svetlana had her camera out and was shooting photo after photo as Brooklyn, Zack, and Dianne lowered the AUV into the water. Svetlana quickly put her camera down, grabbed her GoPro camera, dove into the water, and started photographing Dr. Wood and Sam as the AUV descended.

The AUV's thrusters took it out of sight into the cave. Ten minutes later, Svetlana photographed Dr. Wood and Sam surfacing. Once out of the water, Diane asked Brooklyn, “Weren’t we going to map the cave system up here?”

“Wow, Diane, you are correct. With everything going on, I completely forgot. Zack, do you remember where we put the compass and the 100-foot measuring tape?”

Zack went to ‘Little Red’ and reached into a box. “Here they are.”

The Cave

Starting from where everyone agreed, Diane, Brooklyn, and Zack mapped the distances and directions the AUV recorded the day before using the compass and measuring tape. An hour later, they came trudging back.

“Dr. Wood, we mapped the cave passages. It starts here and goes over to here. Right here, there is a branch that goes off to the right. We marked it with these rocks. The passage with the line continues this way and swerves to the left. We marked it with branches and rocks,” said Diane.

Brooklyn completed the narration and the trek a few minutes later.

“... and here is the end of the line. The strap was seen right here.” As Brooklyn pointed to a spot where an old shoe was lying.

"You guys did a great job. Let's make your marking a little more permanent," said Dr. Wood.

Sam pulled out a bundle of survey flags and a can of spray paint. They made the markings more visible and harder to destroy for the next half-hour.

Dr. Wood said, "Zack, if I remember correctly, you know how to fly a drone."

"Yes, Dr. Wood. When we were analyzing the damage from the hurricane last year, I flew a parrot Bebop drone for about a week straight," said Zack.

"Excellent, well, I have one in that box that I lugged here. I need you to fly over this area. Fly about two hundred feet above us so we can see where the cave tunnel is. While you are at it, teach Diane and Brooklyn how to fly it."

While Zack, Diane, and Brooklyn went off to fly the drone, Dr. Wood, Sam, Svetlana, Arthur, and Sergeant Mitchell sat back and ate lunch. Svetlana's sandwiches were "To-die-for," as Zack had said the day before. Arthur and Dr. Wood reminisced over old times during lunch and caught up on what had happened.

It was 10 a.m. when the AUV dove. At 2 p.m., it came up again. Brooklyn and Diane immediately connected their laptop computer to the vehicle using Bluetooth and downloaded the latest dive data. The downloading took about an hour, and when complete, Diane switched the vehicle to auxiliary power (that is, the generator they had lugged along). Quickly, Zack swapped out the batteries while Brooklyn and Diane sent the command for the vehicle to continue from where it left off. They also set the maximum time limit to 3 ½ hours since it was late. By 3:15 p.m., the AUV was in the water heading into the cave system.

Brooklyn and Diane focused on the downloaded data, and Zack concentrated on the videos he had made with the drone. Dr. Wood, Sam, Svetlana, Arthur, and Sergeant Mitchell watched and waited.

Svetlana noticed that Dr. Wood started looking at his watch over and over. "What's wrong, Dr. Wood?"

"I don't know; I just have this feeling that we need to speed up the search today. Let's immediately get all the unnecessary stuff back on the bus—especially the computers. Brooklyn, please make a backup of this drive. Svetlana, can you take the drive to your boat?" Dr. Wood asked.

"Sure," answered Svetlana.

"Ok, let's get moving."

They all hustled to break down the camp.

Dr. Wood asked, "Brooklyn, can you send an abort mission and return to base to the AUV?"

"Yes," said Brooklyn. "Diane, can you get out the SONAR Transducer? Yes, the big black cylinder in that box."

Diane got it out, hooked up the cable, and lowered it in front of the cave mouth. Brooklyn then hooked the cable to her computer and activated a signal broadcast from the SONAR transducer. A minute later, Brooklyn reported that the AUV acknowledged the abort and was returning at top speed. Half an hour later, the AUV bubbled to the surface just as the team returned from taking their first load to the bus.

Dr. Svetlana spoke up as the AUV was pulled onto land. "Dr. Wood, I can take the AUV on board my boat if you want. I would suggest Brooklyn, Diane, and Zack come with me. We will meet you tomorrow morning."

"That's an excellent plan. Make sure you take the battery charger and Brooklyn's laptop. The pelican case with the dolphins on it," said Dr. Wood.

"Really! We can go with Svetlana! Prima! But what about our clothes and stuff?" Diane asked.

"No problem, I have spare toothbrushes, and we can do a quick wash and dry of your clothes, and yes, there is a shower on board," said Svetlana.

"Then it's settled. Let's get the AUV in the rowboat along with the battery charger and the laptop computer," said Dr. Wood.

Half an hour later, the AUV and AUV support team were on their way to Svetlana's boat while Dr. Wood, Sam, Arthur, and Sergeant Mitchell returned to the bus.

“Svetlana, how big is your boat?” Diane asked as they approached it.

“I have equipped a 65-foot NEEL trimaran for scuba diving. I use it mainly to take tourists on extended trips. I always have extras of about everything, including ties, cufflinks, and shoe polish,” said Svetlana.

As they came to the boat, an older man approached the railing.

“Hi Svetlana. Throw me a line.”

"Hi James. We have company, and we will need the derrick to lift the AUV onto the rear deck,” said Svetlana.

The rowboat was quickly tired off.

“Everybody, this is James, my number one assistant. James, this is Diane, Brooklyn, and Zack.”

Svetlana’s Trimaran

"Hi, everyone. Come aboard and make yourself comfortable. It will take a few minutes to mount the derrick," said James.

They all climbed on the boat and sat back while James went below, and, a few minutes later, he mounted the derrick, slid it into place, and bolted it down. James then rotated it over the site where Svetlana clipped the cable to the AUV. James flipped the switch, and the AUV rose into the air. James pushed the derrick arm and rotated the AUV to the back deck, where he lowered it to the deck. He then took out a hose and sprayed it down with fresh water.

Svetlana took Brooklyn's computer inside, along with the battery recharger, and hooked them up. Zack immediately attached the recharger leads.

"Now for the tour of my boat," said Svetlana.

"The boat can sleep ten. The master cabin is through there," she said, pointing forward. "Brooklyn and Diane, you two will be in the starboard cabin through there. Zack, you will be in the port cabin over there. As you can see, there is the galley (the kitchen). Behind that door is the head, yes, the bathroom. You will find a shower there. Remember, nothing down the toilet that didn't come out of you and the toilet paper supplied! Now, make yourselves at home."

They all looked around and went to their assigned cabins.

You could hear throughout the boat comments like, "This is so neat!" "Look at these!" "I wonder what this is for?" Twenty minutes later, Svetlana said, "OK, time to make dinner. Brooklyn, you are exempt since you need to process all the AUV data. Diane and Zack set the dinner table. Here are some vegetables to cut up. You guys like fish, I hope? Yesterday, James caught a couple of dolphin fish, no, not Flipper, many people call them Mahi Mahi ..."

The radio suddenly turned on … "Hurricane Dorian is forecast to strike the Bahamas islands at a devastating category 5 intensity, driving life-threatening storm surge flooding, destructive winds, and flooding rainfall. Dorian will also spread heavy rain and strong winds to the State of Florida after it passes through the Bahamas islands. Dorian could become a potentially catastrophic event for the Bahamas and mid-eastern Florida. Dorian could be the strongest landfall along the mid-eastern sections of Florida in 10 years.

Dorian is currently centered about 335 miles south-south-east of the Bahamas islands and is moving at 12 miles-per-hour…."

"Damn, you will have to find Rob and Leanne very soon. There isn't much time left. When that hurricane hits, no one will think about them for months," said Svetlana.

As they prepared dinner, they all talked about hurricanes. Only Diane had never experienced one before and asked question after question: Which hurricane was the biggest? Does the island get hit often? What is it like to be in a hurricane? And so on. Brooklyn had been through two hurricanes that had come close to the university, Zack had been in one that went through the Florida Keys, James had been through five on various islands in the Caribbean, and Svetlana through three, one of them on her boat.

"Oh, my, that was one scary day. I'll tell you we were playing dodgeball. The forecast said it was going north, so we went south; then, just as we thought we were safe, it changed directions and came straight at us. We had nowhere to run. Land was on one side but without protection, and a storm was on the other. I kept as much sail up as possible until we cleared the land. This boat was fast, but we were cornered. Our only chance was to put out a sea anchor and ride it out facing the waves. Those waves were huge! I learned more about severe storms and hurricanes in 48 hours than I would have learned in a lifetime at sea. The wind and seas really battered the boat until the mast broke. Luckily, I had put out a sea anchor, which kept the boat aligned with the waves; otherwise, the waves would have hit it broadside and eventually would have broken it into pieces. I hope never to have to sail through another hurricane again."

Diane asked, "Exactly what is a sea anchor?"

"Think of it as an anchor that doesn't go to the bottom. It can be made of anything. I used whatever I could: old sails, clothes wrapped in a canvas bag, and so on. Anything that will help the boat to face the waves" said Svetlana.

The conversation continued until dinner was cooked. It was a beautiful evening, so they sat on the back deck for dinner. Not a cloud in the sky. It was almost as if something had sucked the clouds away. None of the usual thunderstorm clouds were around.

"Wow, what a beautiful evening," said Diane.

"The calm before the storm," said James.

"What do you mean?"

James continued. "You see, a hurricane draws all the weather around it. We have beautiful weather because of the hurricane."

After dinner, they sat back for half an hour until the computer processing the AUV's video dinged that it had finished processing the data. Like the previous evening, they watched the first five minutes at normal speed and then increased the speed. They slowed it down again when the AUV approached the strap. The AUV went deeper and deeper into the cave, and at the end of this first search of the day, there were scratches on the left side wall.

"Wow! Look at that!" Svetlana exclaimed. "That was a hard blow, probably with a scuba tank. Look at these scrapes here. Brooklyn, give me the video picture of what's just above us right now."

Brooklyn switched the view to video – the top view camera showed what was above the vehicle. Svetlana immediately said, "Ah ha, yes, someone was here!"

"How do you know?" Diane asked.

"See these shiny reflections. That is the air that is in the pockets of the cave ceiling. Eventually, they will dissipate. The air in these pockets is very recent."

"So, it is highly likely Rob and Leanne were here and put up a fight. But against whom?"

James broke in on the conversation, "Svetlana, you better come on deck. Something is happening on the mainland."

They all went on the back deck. Shouts came from the area around the Blue Hole. Flashlights lit up the area. It was a very dark night, and none of the people on land had thought to shine a light out to sea. Angry voices were heard, and then a shot rang out. Then they heard, "You bastard, you shot me. Damn, that hurts. Oh god, oh god." Then there was silence.

"Why did you have to go and shoot him? What is Mr. Rodrigues going to say?" Then, they heard some highly explicit language.

Svetlana quickly said, "Turn off the lights right now! James, please turn off the navigation lights, put floats on the ends of the anchor lines, and release them. I will get the motors going as soon as you release the last one. Everyone else get below and lay on the floor. I expect they will shoot at us next if they see us."

Everyone quickly went below. With all the lights out and everyone quiet, they could hear the people on land, most of it in Spanish, some in English, but they couldn't make out what was being said. James silently tied the floats to the ends of the anchor lines and slid them silently into the water until the last was overboard.

Svetlana turned the motor on. To everyone on the boat, it sounded like a roaring engine, even though it was actually a light whisper of a motor.

Diane looked over at Svetlana as she motored the boat carefully away from land. Diane gasped as she saw what looked like an alien at the wheel! Svetlana had protruding lenses sticking out of her eyes. Diane whispered to Brooklyn. "OMG look at Svetlana! She is an alien. What are we going to do?"

Brooklyn looked at Svetlana and had to hold back her laughter. "Diane, those are night vision goggles."

"Oh, yes. I knew that," said Diane shyly.

They glided further and further away from shore and were soon out of sight from the people on land. Everyone came to the bridge.

Zack promptly started, "Now that was a close one. Those were definitely drug runners. A few years ago, the Coast Guard encountered a sailboat with everyone dead on it. It was drug runners who used the boat to ship a load of drugs into the States. The Coast Guard was surprised to find the boat since normally they sink the boats they use."

"So, what do we do now?" Diane asked.

"Find a place to anchor for the night that is hidden and unlikely to be known by foreigners," said Svetlana.

An hour later, they anchored in a little hidden cove. Svetlana tried to call Dr. Wood on her cell phone, but she only got his answering machine. She then left a message on Arthur's answering machine telling him what had happened and that someone should check up on Dr. Wood and Sam back at the hotel.

"Brooklyn, have you finished reviewing the AUV data?" Svetlana asked.

"No, but I will get on it immediately," said Brooklyn quickly.

They all returned to the computer and continued looking at the AUV video.

The AUV continued for another fifty feet and became much wider when suddenly ahead, they saw the side of something directly in front of them. It was a mini-sub going by an intersecting tunnel! A light shown directly at them, and an eye peering out from an observation pothole.

"Oh, damn," said Svetlana. "They know we are here."

"What are we going to do?" Diane asked.

"Good question. We are safe right now, so let's get some sleep, and I'll come up with something. Everyone, off to bed. It is getting late, and I think tomorrow will be a long day," said Svetlana.

Everyone headed to their staterooms. Zack stopped at his door, turned around, and returned to the computer. Turning it back on, he got out the USB drive with a copy of the drone video. As soon as the computer was on, he started viewing the drone video, looking for anything unusual, such as objects that might look like an entrance, an obvious path leading to nowhere, or a small building. Zack looked and looked, but there was so much junk everywhere. He found five potential sites, printed out overall views and close-ups of each of the five, and then knocked lightly on Svetlana's door. Thirty seconds later, it opened with a worried look on her face. Svetlana asked: "What is it?"

"Svetlana, sorry to intrude, but just as we headed to bed, I thought of all the aerial video footage we took with the drone. I immediately started reviewing it to see if I could find anything related to the drug traffickers. I think there is some underground facility where the submarine docks. If there is a facility, an entrance must be nearby. The question is, where and what does it look like? First, I started with the view covering everything from where we parked the bus to way over here." Zack pointed to the edge of the overall picture he had printed. "Take a look at this spot near the road. It is a small pickup truck hidden behind some coral reef blocks and brush. I think that is their vehicle to travel around the island."

"Excellent Zack," Svetlana said as she viewed the picture closely. "What else have you found?"

"Well, nothing definite. I came up with a list of things that would help locate an entrance, such as new trails, areas with tracks, areas with a lot of trees or high brush, and so on," said Zack.

"And?"

"Well, I found five possible locations using my criteria. This looks like a small mobile home or trailer. There are tracks or what could be tracks. Look at this pic," said Zack.

"Yes, that is a possibility. Nice picture," said Svetlana. "What next?"

"Here is a flat rectangular plate with a sandy patch right here as if it had been scuffed by feet."

"Yes, it does indeed," said Svetlana.

"The third possibility is the object in this photo. It might just be a pile of rocks, but it might not," said Zack.

Svetlana looked closely. "I see what you mean. It definitely is a spot that needs to be checked."

"The fourth possibility is right here in this photo. It looks like stairs going down. It is at the edge of the drone videos, so it might be stairs going up, like at the top of a tower.

Svetlana stared at it. "Yes, you are right; it is probably a tower. So, what do you have for the fifth possibility?"

"Right here, look at this photo. A curvy line of high brush and trees runs from their pickup truck to here and eventually to this rusty, derelict VW Bug. Except for these white marks here, nothing is suspicious about it."

Svetlana looked intently at the photograph: “Excellent Zack. I will take a good look at each tonight. Get a good night’s sleep. I think you are right. One of these sites is their entrance. We will inform the authorities tomorrow when they get here.

FOLLOW THE UNKNOWN TRACK

6 a.m. Saturday, August 16th

"Hello Dr. Wood. What is your status? Speak louder; there is a lot of static," said Svetlana over the boat's radio.

"We are all fine; the bus has been destroyed. It was blown up just before dawn. The room next to ours was shot up at the same time. It looks like they tried to kill us. How is everyone there?" Dr. Wood asked.

"We are all fine. During dinner, we heard shouting, and then we heard a shot. It looks like they killed one of their own. We quickly up anchored with lights out and moved to a safe anchorage." said Svetlana.

Dr. Wood continued: "It looks like they blew up the cell phone towers on the island. Arthur, Sergeant Mitchell, Sam, and I are bringing the island police to the Blue Hole."

"Dr. Wood, we think Rob and Leanne were captured. On the AUV's video recording, we found evidence of a fight. There were no bodies. Then, a tourist-type submarine went by, and they saw us, which is why I think they did what they did last night; there must be an entrance somewhere in the area," said Svetlana.

"Perfect," said Dr. Wood. "Get back to the Blue Hole. We will meet you there. I need to send the AUV in again to watch and maybe flush them out," said Dr. Wood.

"OK, we will head back. We will continue to monitor this channel. I recommend that the police send a boat and that the American Coast Guard be notified," said Svetlana.

"Roger that. Over and out," replied Dr. Wood.

Svetlana promptly said in a stressed voice. "Anyone for breakfast? Pancakes, bacon, and eggs with pure maple syrup, OJ, or coffee. Brooklyn, have you recharged the AUV's batteries?"

"Yes, they are all charged up, and the vehicle is ready to go, and yes, I'd like pancakes," said Brooklyn with a smile.

"Me too, me too," replied Diane and Zack.

Svetlana and James raised the anchor as soon as they were done and headed to the cave just off the beach from the Blue Hole. They were arriving when a police boat with its lights flashing zoomed up to them.

Arthur waved from the bow. "Svetlana, fill us in. What do we know?"

"Hi Arthur, we are so glad to see you."

Svetlana filled them in on what she knew and what had happened and handed over Zack's photos. "Here are five places you can start your search."

"Excellent Svetlana," as he viewed the pictures. "Thank you. You go in first and pick up your anchor line, and we will come alongside. When we go ashore, I expect Sergeant Mitchell, Dr. Wood, Sam, and several police to arrive," said Arthur.

Svetlana proceeded to cruise to the exact location, oriented the boat as before, and set out fenders as the police boat came up and tied off to the trimaran.

The police went first to secure the area. Fifteen minutes later, Arthur radioed that everything was safe and that they could come ashore.

Svetlana took charge. "Brooklyn, Diane, Zack, and Dr. Wood will need the AUV immediately. We have two things going against us. First, we have a hurricane bearing down on us; if you noticed, the first clouds are here, and the winds are beginning to pick up; and second, the drug traducers know we are on to them and will be trying to escape and may kill Rob and Leanne in the process. So, we must hurry. Brooklyn, how did you program the AUV for its next mission?" Svetlana asked.

"I set it to continue its search. If it finds any manmade object, it is to be documented. If it's really big, it must be documented and returned immediately. If it is a body, document it and return immediately. If anything attempts to grab it, the AUV will take evasive action and put out a very loud, debilitating sound pulse. If captured, it will continue to put out extremely load pulses until the battery dies. Documentation will continue until the computer has to shut down for lack of power," replied Brooklyn.

"Excellent," said Svetlana. "Let's get the AUV on the Zodiac!"

Brooklyn, Diane, and Zack carefully lowered the AUV into the boat. Everything was loaded, including a cooler with drinks and sandwiches. The boat ride was quick, and soon, they unloaded the boat. With the help of two police officers, everything was carried to the Blue Hole.

At the hole, they immediately started running through the preflight check they had done a hundred times before. As they completed the checklist, Dr. Wood, Sam, Sergeant Mitchell, and ten other police officers walked up. There must have been almost every police officer on the island there.

Arthur filled them in that they were starting the search for an entrance to a suspected underground facility using the photos supplied by Zack from the drone survey.

When Diane overheard this, she gave the thumbs-up to Zack with a big smile.

"Brooklyn and Diane, I heard from Svetlana that the vehicle is ready and that your missions are well planned. Launch it when you are ready," said Dr. Wood.

"Brooklyn," said Diane, "all ready."

"OK, Zack, grab a corner. Let's lower it in the water," said Brooklyn.

Each grabbed hold, and the three then lowered it down. The moment it was entirely in the water, Svetlana dove into it, following the vehicle down. Soon, the AUV was out of sight, and Svetlana was climbing back on the bank, giving the thumbs-up that everything with the vehicle was okay.

Minutes later, one of the police reported that the hidden pickup truck had been found and that they had radioed back to find out who the owner was.

A half-hour later, a shout reported that a passageway was found in the shack Zack had suggested as a likely entrance point—five minutes later, another shout. There was an opening under the VW Bug. They also found stairs going down under a rectangular metal door.

"Wow Zack," said Diane. "You found them! Sweet!"

Guards were placed at each entrance while everyone gathered to formulate a plan. All of this took until noon.

Brooklyn approached Dr. Wood and Sam. "Dr. Wood, I'm concerned about the AUV; it should have returned ten minutes ago. Should I send out an abort signal?"

Just as Dr. Wood was about to respond, everyone heard it like an air-raid siren from World War II or emergency warning speakers informing that a tornado was nearby; the AUV was putting out a distress signal.

"Oh, no! The AUV has been captured and is out of the water," cried Diane

The police made their final arrangements to enter all three entrances simultaneously. Overhearing this, Diane noticed thick clouds covering the sun, and a breeze blew through the trees and bushes. The wailing of the AUV stopped abruptly.

The radio broadcasted that the police were entering the facility. An hour later, it was reported that five people were captured …

An hour later, the police reported that the submarine had escaped and that they had recovered hundreds of pounds of cocaine. Also, there was a room filled with empty scuba tanks.

"And what about Rob and Leanne?" Dr. Wood asked.

Sergeant Mitchell came up and said, "No sign of them. There are two sets of scuba gear, one men's and one women's. They were lying in a corner. Oh, I'm sorry to say your AUV took some damage. They couldn't figure out how to turn off the siren, so they just started beating it until it stopped. You can go down and look for yourself in a few minutes."

Fifteen minutes later, they entered the underground facility by the VW-Bug entrance. The drug traffickers had put a hydraulic system to lift one side of the car by pushing a button attached to a nearby tree.

It was damp and clammy until they passed through a door at the bottom of the stairs. Beyond the door, it was dry and comfortable. Someone had put a lot of money and effort into making the place a livable and working facility. One of the police officers guided them to the underground harbor.

Dr. Wood went to the scuba gear. "Yes, this is Rob's. He always puts his name at the bottom corner: see right here!"

"I'm sorry. They can still be alive on the submarine," said Svetlana. She then continued. "Dr. Wood, did you ever give your Nautilus team a call?"

"I forgot to tell you, so much has been happening. Yes, it is finished and not tested but finished. I told them to get it over here ASAP. They will be here ... what day is it?" Without waiting for a reply, Dr. Wood continued. "It arrived an hour ago at the airport unless there were delays."

"Dr. Wood, get your Nautilus team here ASAP. Ask Sergeant Mitchell to use his police radio. We can load everything on my boat. I have a multi-beam SONAR system on the bow of my boat; we may be able to find them," said Svetlana.

Dr. Wood headed to the surface with Sergeant Mitchell and Sam to make the call.

A few feet away, Diane asked, "Brooklyn, how's the vehicle?"

"Bad, really bad." Tears were flowing down her cheeks. "Just look."

Diane looked. Yes, it was in a bad way. Someone had used a sledgehammer, so it looked like an OREO cookie.

"Now we know that the OREO cookie is the perfect name for it," said Diane.

"What do you mean?"

"It sure looks like an OREO cookie now," said Diane sorrowfully.

Zack laughed and pulled on Diane's arm. “Diane, we must return to Svetlana's boat right now!”

“Why?”

"Things are going to happen fast. The hurricane is almost here, Svetlana needs to get her boat out of here, Dr. Wood needs to locate the sub, we have to rescue Rob and Leanne, and they won't want us to be there because we are just kids in their eyes.”

“OK, let's leave. Brooklyn will be of no use. She is crying too much,” said Diane.

THE CHASE

As they passed into the passage, they heard Svetlana telling Dr. Wood to meet her back at the boat. Quickly and quietly, they returned to the surface.

"How are we going to get to the boat?" Diane asked.

"Simple, we just ask one of the police officers to help us out. No one will suspect a thing," said Zack.

Sure enough, a policeman helped them in the Zodiac, zoomed them to the trimaran, and brought them back to land. James had come out to help them on board. Diane immediately explained that Dr. Wood and Svetlana would be coming back on board and would give chase to find the submarine. Diane also told him about the underground facility, how they found the scuba gear, and that the AUV was severely damaged.

James started preparing the boat for an immediate departure while Diane and Zack went below to make themselves scarce. When adults want to do things fast, they want to avoid children or even young adults potentially getting in the way.

An hour later, they heard shouts and noise on deck. Words like, "Get it on board," "Put it there," "Get the anchors up," "Tie off the Zodiac for towing," and so on. People were soon in the main cabin, motors were started, and they could feel the boat begin to move.

"This is almost like stowing away," said Diane with a smile.

"You must try to hide to be a stowaway. We are just hanging out in our beds," said Zack with a devious smile.

The door suddenly opened, and Svetlana stuck her head in and said with a grin. "Come on deck. We are leaving. Dr. Wood thought you two might be interested in the Nautilus project."

"What, you knew we were here," said Diane.

"Of course, you don't expect your dad not to keep an eye on you. I gather you are famous for getting into places and situations that are … let's say questionable," answered Svetlana. "Then, of course, a policeman let us know you two were onboard, and James told us how you got him ready for a quick departure. Thanks, by the way. It did speed up the process. Good thinking. Now come on deck and see Dr. Wood's latest contraption."

Dr. Wood and two of his graduate students were on deck working on what looked like a hybrid plane-boat-sub. It was the NAUTILUS, an autonomous underwater vehicle (AUV) – autonomous surface vehicle (ASV) – unmanned aerial vehicle (UAV). It was designed to 1—cruise on the water's surface, 2—Be able to take off, 3—Fly, 4—Land on water, 5—Dive into the water, and 6—Swim underwater.

"What do you think, Diane?" Dr. Wood asked. "This vehicle dives to 100 feet and swims at two and a quarter to two and a half miles per hour. The vehicle returns to the water surface to take off. It cruises 35 miles an hour. Right now, we have a camera to record the surroundings and various sensors to analyze the vehicle's position, such as air pressure and location via GPS and multiaxial gyroscope sensors."

"Wow!" said Diane with her mouth open

"Oh yes, let me introduce Brian and Tij. They have worked hard to get this ready. They are quite upset that they did not have more time to test it. Well, it will be tested today!" said Dr. Wood.

Arthur, Sergeant Mitchell, Sam, and the two police officers were also on deck. James was on the bridge, guiding the boat into deeper water. Svetlana headed to the bridge to turn on the multibeam SONAR, which would give them a nice acoustic picture of the submarine below if they came upon it.

A little after 3 p.m., Dr. Wood asked, "Svetlana, do you have any of your wonderful sandwiches?"

"Oh my, in all the excitement, I forgot I had made ten sandwiches, and plenty of drinks are in the cooler. Did it get back on board?"

Arthur pointed to a corner spot out of the way. Zack went to the cooler, grabbed them all, and passed them out - twelve people, ten sandwiches. Who will be the ones without a sandwich? When he came to James, James said, "Thank you very much, except I already ate. That left Dr. Wood, Sam, Diane, Svetlana, Arthur, two policemen, Brian, and Tij. Zack passed out the last sandwich. Oh, drat! He was the one without. At that moment, Svetlana returned to the rear deck and gave him a freshly made sandwich.

"See, Zack's good deeds can be rewarded. You will really like this sandwich," said Svetlana.

And boy, was she so right. It was fabulous!

In the middle of lunch, Dr. Wood asked everyone. "Does anyone know where the exit tunnel might be, in which direction they might go, and why?"

Everyone paused for a moment and then continued eating at a much slower pace. Diane raised her hand.

"Diane, this is not school; you don't have to raise your hand," said Dr. Wood. "What do you have?"

"I have two suggestions. 1) Call on the radio for any scuba diver who dives in this area. Ask if they have any ideas; and 2) if any fishermen had noticed anything such as a very good or poor place to fish."

"Excellent idea," said Dr. Wood.

"I'll get right on it," said Svetlana. "Arthur, I'll need your assistance."

The two went to the bridge to make some calls.

"Great start. Any other ideas?" Dr. Wood asked. "Especially as to where they might go."

Sam started by pulling a map out of the Bahamas and the east coast of Florida. "All right, everyone, take a good look at this map. It is a typical navigation and dive map with depths in meters. I want everyone to write five locations that might be a possibility of where they would go and write why that is a possible choice. If anyone has a question, we will all try to answer it. Unfortunately, the cell towers are still down, so we have no internet. We have a radio, so if you think someone else may know the answer, we can call them. Here are pens and paper. All right, get started. In 30 minutes, we will see how everyone is. By the way, I want you to guess the likelihood of your suggestion based on your guesses."

"Bayesian statistics is a theory where probability expresses a degree of belief in an event, which can change as new information is gathered. The degree of belief may be based on prior knowledge about the event, such as the results of previous experiments, or, in our case, on personal beliefs about the event. Bayesian statistics uses Bayes' theorem to compute and update probabilities after obtaining new data. Can you believe it? He did this in 1763! One of the most famous cases of using Bayes theorem was in search of the submarine Scorpion, which was heading back to the USA in 1968 when it disappeared off the coast of Spain."

"This guy John Craven, who died a couple of years ago, got all types of experts together, such as mathematicians, submarine experts, and torpedo specialists. He had each of them bet on why the submarine ran into trouble, its speed as it headed to the ocean bottom, the steepness of its descent, and so forth, including why the submarine might have made such a maneuver. He then took the guesses and used Bayes' theorem to estimate where the Scorpion might have sunk. The Navy spent six months searching for the submarine using traditional search methods. Ultimately, they finally agreed to allow Craven three weeks to use his techniques to search for the submarine. A week later, Craven found the sub 220 yards from where he projected it to be! Everyone thought Craven was a lunatic until there it was!"

Forty-five minutes later, after answering many questions, Sam had five possible locations for each person on board. They had tried to call a few people by radio but with no luck.

"Ok. Thanks, everyone. This will take 30 minutes, perhaps an hour, to enter all the data." Sam headed for the computer that was designated for the AUV. Brooklyn, Diane, and Zack followed, curious about what Sam would do.

The very first thing Sam did was to take a thumb USB drive out of his pocket, connect it to the computer, and then turn the computer on. After Brooklyn entered her password, the system booted up, and Sam ran an installation program on the thumb drive. Five minutes later, Sam clicked on a new icon on the computer labeled Craven's Dream.

After a minute or so of running, it came up with a dialog box: What are you searching for? A) Direction, B) Location, and c) Generic solution.

Sam typed "B" and began to enter all the locations. Then, all the confidence levels they had in each area, that is, the likelihood that a specific site would come up. Then he sat back, read through everything, and made numerous notes that were hieroglyphics to the others. Twenty minutes later, he entered the final data sets and announced happily. "Ok, here goes!"

And, of course, as in most programming exercises, the first thing the computer replied was "Error – Data format error. Please check your data."

Sam countered with, "Damn."

After another twenty minutes, Sam was ready. The program ran for thirty seconds and came out with Freeport in the Bahamas, a page of symbols and numbers, and a listing of what the symbols and numbers meant.

"Dr. Wood," called Diane out the door. "I think my dad has found a location."

Dr. Wood came to the computer. "Sam, now that is sweet. How did you do that?"

"Well, it is rather like magic. With all the information everyone provided and what you all thought of the probability that they would be at any one location. I entered the data into a program that uses some of Bayes' theories, and presto, there it is."

"Now, let's plot a course from here to Freeport. The question is, 'will they be going to the north or south side of the island?'" Dr. Wood asked.

"Well, the Grand Bahama International Airport is on the north side. There is also the company Freeport Aggregates just east of there. What do you think?" asked Sergeant Mitchell.

"What about the housing development just west of the airport? There are many channels to lose themselves in. I have a friend living there. I think it is called Queens Cove," said Arthur.

"No, I don't think so; that is a housing community. Everyone watches everyone in housing communities. Just like at home, everyone is a busybody," said Diane, looking up with surprise at the boldness of her statement.

Svetlana approached the map of Freeport and said, "I'd come in right here and dock at the Sir Charles Yacht Club. Many rich people dock there, and I remember in the past that some submersible companies used that dock. That's where I'd go if I were to dock at Freeport."

"Yes, I agree," said Dr. Wood. "Now the question is whether they will go north or south."

"North, of course. Deeper and darker water with fewer islands to go around. It is simpler, and there is more ship traffic to hide under. Then, of course, they probably ran that route hundreds of times to any of the dive locations from West Palm Beach to Miami," replied Svetlana.

"Now that is great teamwork," said Sam. "I wish I could get my team at work to function so smoothly."

Dr. Wood looked at the map. "OK, now, where do you think they are right now? First, when do you think they left? The police entered the facility at 1 p.m. Sergeant Mitchell, any ideas?"

"When we arrived at the underground docking station, it was 3:30 in the afternoon, and the place was empty. So, for sure, it was before then."

"OK, we will assume they left at 12:45. How fast can it go? Actually, how fast do you think they will go?" Dr. Wood asked.

Everyone looked at each other. No one had experience with submersibles. Dr. Wood continued. "Well, I was on the Johnson Sea Link a few years ago. Its maximum speed was only two knots, but submersibles are slow. Let's assume that they can cruise at four knots. A knot is 1.15 miles per hour, so 4 knots is 4.6 miles per hour. Let's say they start off quickly, cruise there, and come in slowly. Let's assume they went at 2 miles an hour for the first hour as they negotiated the shallow waters around here and then sped up to 4 miles an hour. Let's also assume that they are going to take the straightest route. It has now been four hours since they left. So …"

Sam stepped in with a calculator and pointed to the map. "Right about here and in an hour right about here."

"Perfect! Svetlana, can you have James steer an intersecting course? We may not be right, but it is the best we have."

"Yes, let me take this up to James, and we will plot an intersecting course at the fastest speed possible. Sergeant Mitchell, perhaps you will want to supply backup. It is a gamble that this is a wild goose chase, but it would be best if someone were there. Also, everyone, have you looked at the weather? It does not look so good. We will have only this one attempt before I must get this boat to safety."

Svetlana, Sergeant Mitchell, and Arthur went up to the bridge to plot the course and contact the authorities at Freeport.

That morning, the copper color of the sky had given way to a dull and sickly yellow that was washed with an angry red as high clouds thickened, and the winds came up, steadily freshening as the hours passed. By evening, a vertical wall of light, medium, and dark grey, very evil-looking storm clouds rose high into the sky to the east. Periodic gusts blew hard across the island. The waves were quickly building. It was a matter of hours before the hurricane hit, and the possibility of capturing the drug runners and rescuing Rob and Leanne would be gone.

THE CAPTURE

10 p.m. Saturday, August 16th

It was a dark and stormy night, as the cliché said. The waves were still small on the west side of Abaco Island, but the wind was building. It was very dark when they arrived, where they suspected the submarine to be. The marine band radio announcer had been continuously broadcasting the weather data and warned everyone at sea to flee to a safe location. The hurricane was now a category 3 storm with an anticipated 12-foot storm surge. People on the island were to either evacuate or be sure they were in a very strong building known to survive category 3 or higher hurricanes.

"Svetlana, do you see anything on your multibeam sonar?" Dr. Wood asked.

"No, but there is so much area to cover. Is there any chance you can get the Nautilus to listen with its hydrophones for the submarine?" Svetlana asked.

Brian replied: "We have programmed it to listen for all known tourist and research subs. Even if they have modified their submarine, we can detect it using my neural-network identification program."

"What is a neural-network identification program?" Svetlana asked.

"Think of the program as the neurons in your brain. The program takes in all the information it receives from its sound sensors, i.e., hydrophones, and matches it with patterns stored in its memory. By comparing the two, just as your brain would, I can tell you if it is our submarine and in what direction it is."

"Brian, Tij, let's get the Nautilus in the air," said Dr. Wood.

"OK, Dr. Wood. Shall we run remotely operated or autonomously?" Brian asked.

"Can you set it up to run an autonomous grid pattern starting from here on the map and having it travel to this point here? Use a spread of 100 meters and have it stop every 100 meters. Also, set it up so that if the hydrophones detect a propeller's noise that is not ours, it will immediately radio us."

"No problem; Tij just took a hydrophone sampling of our propeller sound and has uploaded it to the vehicle's memory. Using the neural network algorithm that I developed for my Ph.D., we should find them if they are anywhere in range," said Brian.

Twenty minutes later, the Nautilus was launched to search for the submarine. Meanwhile, the trimaran continued in the direction of where they thought the sub would be.

Two hours later, they received a signal from the Nautilus. Submarine below! To the northwest of their current location. Svetlana turned the trimaran in that direction. The light on the back of the Nautilus became brighter.

"Dr. Wood! There he is! I have him on our multibeam SONAR!" yelled Svetlana from the bridge.

"Sweet, get Arthur and Sergeant Mitchell to the bridge," said Dr. Wood to Zack. "Everyone to the bridge."

Two minutes later, everyone was looking at the image on the screen. The submarine was highly visible. Svetlana said, "Well, there they are, and they will only be there for about half an hour more."

"Why do you say that?" Dr. Wood asked. "It looks like we have them."

"Not really," continued Svetlana. "Take a look right there. That is the edge of the shallow zone. The sea floor drops very deep, to around 2000 meters deep. I don't know how deep that submarine can go, but much deeper than this multibeam sonar can penetrate."

"Damn! Do you mean that we will lose them right there?" Arthur asked.

"Yes, there is no way to stop them even when the police boat and coast guard arrive, which should be soon. Unless they have depth chargers, which I have many doubts about," said Svetlana.

"So, what will we do?" asked Sergeant Mitchell.

Diane spoke up from behind everyone: "Is there some way we can use the Nautilus to stop them? Perhaps have it jam their controls."

"Hmm, that is a possibility. Perhaps wrap a wire around their rudder. Tij and Brian, what do you think?" Dr. Wood asked.

"No way. We don't have such fine control to do such a thing. If we have a pinger, we might be able to attach it to the hull, and then we could track it wherever it went," replied Tij.

"Now, that is not something we planned for. Svetlana, do you have anything on board that can act as a pinger?" Dr. Wood asked.

"That is a good question," replied Svetlana. "James, do you know of anything?"

Everyone pondered for a minute. Then, James began very slowly. "There is a possibility. Svetlana, remember five years ago, we helped a Florida Institute of Oceanography scientist. He was studying fish and had these pingers to ward off sea mammals. He left one of his nets. I meant to return a few years back and forgot it was there. Perhaps there are some pingers attached. I'll go into the storage hold and check." James went below to the forward storage hold.

Five minutes later, he was back on deck with five pingers and handed them to Svetlana, who gave them to Dr. Wood, who handed them to Sam, who handed them to Diane. Diane looked at them and asked, "What am I supposed to do with these?"

"You have 10 minutes to get one of them to work and another 5 minutes to attach them to the Nautilus with Brian and Tij. Then they have 15 minutes to get the Nautilus to the submarine to attach the pinger," said Dr. Wood. "Everyone, GO!"

Zack, Diane, and Sam grabbed the pingers and immediately broke one open with a sharp knife. "Yes, the battery is long dead. It takes two AAA lithium batteries. Does anyone have any?" Diane asked. "Also, we need some glue. Epoxy or Shoe Goo or anything!"

James came forward with five-minute epoxy. "This should do the trick unless it is too old. I bought it for a couple of repairs a year ago."

Sergeant Mitchell opened his walkie-talkie and handed over four AAA lithium batteries. "Just replaced them the other day, so they are fully charged."

"Excellent, that means we can get two pingers going. Tij, Brian, prepare the vehicle to carry two," said Dr. Wood.

"Ah, Dr. Wood, the vehicle isn't designed to carry so much. It just won't work!"

"Oh yes it will," demanded Dr. Wood. "Dump some weight off the vehicle. Look here. Chop this off, and this and this."

"Dr. Wood, we won't be able to take off again if we do that!"

"We won't need to. It is way more important to get down there and to attach the pingers than to get the vehicle back," said Dr. Wood.

Everyone got to work. Two pingers were diligently cut open, and the batteries replaced and resealed with the five-minute epoxy. Tij and Brian attached the two pingers to the midsection of the Nautilus so it wouldn't change the center of gravity (an imaginary point in an object where the body's total weight may be thought to be concentrated).

Fifteen minutes after they started, they had two pingers attached to the Nautilus, which Brian launched, and Tij guided them to a spot in front of the submarine.

"Dr. Wood. This was a rush job. I only had time to tell the vehicle to crash into the submarine. We attached very sticky stuff to the pingers so that when contact is made, they will attach themselves to the submarine," said Tij.

"Excellent. Now we can keep our fingers crossed," said Dr. Wood.

"Everyone, to the bridge. We can watch the action on the multi-beam sonar," said Svetlana.

They all quickly found seats, looking at the multibeam sonar monitor. The bottom contours were all visible in the foreground, with a dark area at the top moving down. "What is that?" Diane asked.

“That’s the drop-off. You can see the submarine right here. It is almost to the dark zone. Now look! Right there! The Nautilus just entered the picture,” said Svetlana.

The Nautilus was speeding to the submarine. The submarine disappeared into the black zone with one hundred yards to go. Then, the Nautilus disappeared.

“Well, Brian and Tij, whether we capture them is now up to your programming. Let’s say you will be heroes or ….” Dr. Wood’s voice trailed off, then pointed at the screen. “What is that?”

Brian replied. “That’s what is left of the Nautilus. It was my last-ditch effort to stop the submarine. I programmed that if it failed to attach the pinger to the sub, it would drop into one of the horizontal thrusters. There is a titanium support rod in the middle of the vehicle. I figured that it would stop the propeller from turning. That way, the submarine could only go around in circles and would eventually have to surface. If the Nautilus did its job, the submarine should be on the surface when the police boat is here.”

“You sacrificed your love and joy! Oh, my goodness,” said Dr. Wood.

Twenty minutes later, just as the police boat approached them, the submarine surfaced. With spotlights on them, the people on the sub came out of the hatch. First, Leanne, then Rob, Rodrigues, two of his henchmen, Miguel and Carlos, and finally, the submarine pilot Luke. The rear right thruster was all tangled in a small titanium rod. The Nautilus had successfully completed its first and last mission.

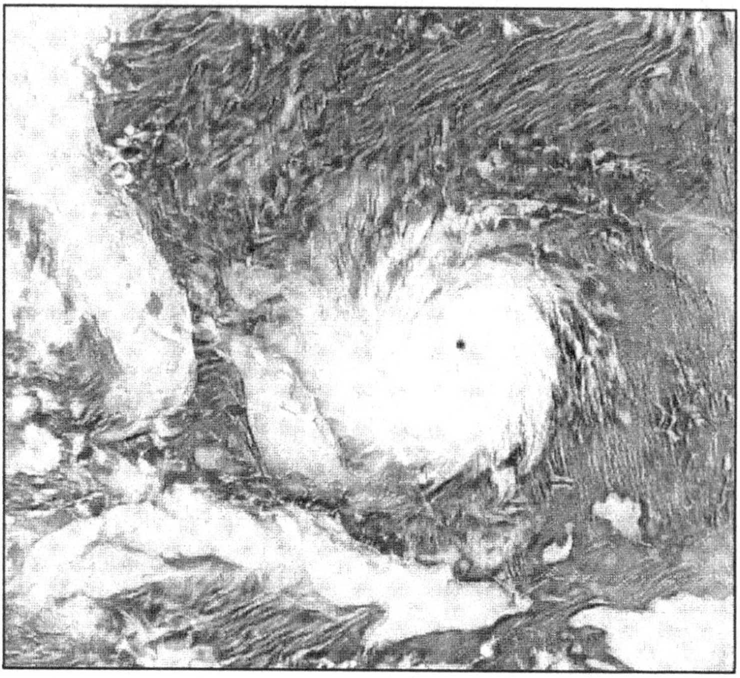

HURRICANE DORIAN

1:30 a.m. Sunday, August 17th

As lines were fastened to the submarine, the winds picked up considerably, increasing gusts up to 50 miles per hour. Svetlana yelled, "Everyone except for James, get off my boat now! Your best bet is to go with the police boat. Arthur, Sergeant Mitchell, I would suggest skip towing the sub back to the dock; you don't have the time! Put a couple of large marking floats on it and come back for it after the storm. Perhaps it will still be on the surface. If not, nothing is lost. You have enough evidence to put these drug runners away for a long time."

"You're right," replied Sergeant Mitchell. "I like your suggestion about tying off the submarine to floats. Everyone who's going ashore onto the police boat … NOW!"

Dr. Wood, his graduate students Diane, Zack, and Sam grabbed the few items they had and jumped onto the police boat, which was now having difficulty holding position due to the size of the waves. When they were off the trimaran, Svetlana put the helm over and sped into the darkness to find a safe place to weather the storm. With a blink of an eye, they were gone.

The police boat was fully loaded with all the extra people on board. The captain put the engines in gear and steered toward Freeport Harbor, several miles away. The police medic was tending to Rob and Leanne. They had been beaten and would need to be checked out at the Freeport hospital for broken bones and internal injuries.

The ride was becoming more and more bouncy. The winds were now into the hurricane range of 75 plus miles per hour and were expected to rise to 175 miles per hour with gusts to 200 miles per hour. The waves were getting huge and choppy, bouncing the boat around like a toy. Finally, they entered Freeport's harbor. At the police dock was an ambulance for Rob and Leanne, a couple of armed security guards to take the drug runners off to prison, and a van to take the other civilians to a hurricane shelter. As horizontal rain and one-hundred-mile-an-hour winds rocked the van back and forth, they crept to the shelter. Luckily, the entrance was shielded, and they could make it inside without incident. As the van slowly drove to its next assignment, Dr. Wood and the others entered a hall filled with people. A tall man came over and showed them a couple of cots that were still empty at the far end of the hall and pointed out where the facilities were and where they could get a bite to eat.

"Well, everyone. We did it," said Dr. Wood with a sigh. "We rescued my friends. I want to thank everyone! You were all magnificent! Each and everyone went way beyond the call of duty. I cannot express my pride in having such a great team. And you, Diane, and you, Zack. Never have I met two kids with such spunk, abilities, and talent. My God, I want you to come and work with me on my next projects. Look me up if you ever need anything." With that, he gave both big hugs.

---- The End ----

Buoyancy

For argument's sake, a soccer ball is found floating in the ocean and the ball has a mass of 0.5 kg and a diameter of 22 cm. a) What is the buoyant force? b) What is the volume of water displaced by the ball? c) What is the average density of the soccer ball?

(a) To find the buoyant force (F_b), simply draw a free-body diagram. The force of gravity is balanced by the buoyant force:

$\Sigma F = ma$

free-body diagram

$F_b - mg = 0$

$F_b = mg = 4.9$ Newton(N) $1N = 1$ kg·m/s

(b) By Archimedes' principle, the buoyant force is equal to the weight of fluid displaced.

$F_b = \rho V_{disp} g$

$V_{disp} = F_b/\rho g = 4.9$ kg·m/s $/(1000\text{kg/m} * 9.8\text{m/s}^2) = 5 \times 10^{-4}$ m^3

(c) To find the density of the ball, we need to determine its volume. The volume of a sphere is:

$V = (4/3)\pi r^3$

With r = 0.11 m, we get:

volume of soccer ball = $V = 5.58 \times 10^{-3}$ m^3

The density is mass divided by volume:

$\rho = m/V = 0.5 / 5.58 \times 10^{-3} = 90$ kg/m^3

Another way to find density is to use the volume of displaced fluid. For a floating object, the weight of the object equals the buoyant force, which equals the weight of the displaced fluid.

$mg = F_b = \rho_{fluid} V_{disp}\, g$

$m = \rho_{object} V$, so:

$\rho_{object} V = \rho_{fluid} V_{disp}$

Factors of g cancel. Re-arranging gives, for a floating object:

$\rho_{object} / \rho_{fluid} = V_{disp}/V$